MATERIALS SCIENCE AND TECHNOLOGIES

PHASE CHANGE MATERIALS

TECHNOLOGY AND APPLICATIONS

MATERIALS SCIENCE AND TECHNOLOGIES

Additional books and e-books in this series can be found on Nova's website under the Series tab.

MATERIALS SCIENCE AND TECHNOLOGIES

PHASE CHANGE MATERIALS

TECHNOLOGY AND APPLICATIONS

ISMAËL VAN DER WINDEN
EDITOR

Copyright © 2020 by Nova Science Publishers, Inc.

All rights reserved. No part of this book may be reproduced, stored in a retrieval system or transmitted in any form or by any means: electronic, electrostatic, magnetic, tape, mechanical photocopying, recording or otherwise without the written permission of the Publisher.

We have partnered with Copyright Clearance Center to make it easy for you to obtain permissions to reuse content from this publication. Simply navigate to this publication's page on Nova's website and locate the "Get Permission" button below the title description. This button is linked directly to the title's permission page on copyright.com. Alternatively, you can visit copyright.com and search by title, ISBN, or ISSN.

For further questions about using the service on copyright.com, please contact:
Copyright Clearance Center
Phone: +1-(978) 750-8400   Fax: +1-(978) 750-4470   E-mail: info@copyright.com.

NOTICE TO THE READER

The Publisher has taken reasonable care in the preparation of this book, but makes no expressed or implied warranty of any kind and assumes no responsibility for any errors or omissions. No liability is assumed for incidental or consequential damages in connection with or arising out of information contained in this book. The Publisher shall not be liable for any special, consequential, or exemplary damages resulting, in whole or in part, from the readers' use of, or reliance upon, this material. Any parts of this book based on government reports are so indicated and copyright is claimed for those parts to the extent applicable to compilations of such works.

Independent verification should be sought for any data, advice or recommendations contained in this book. In addition, no responsibility is assumed by the Publisher for any injury and/or damage to persons or property arising from any methods, products, instructions, ideas or otherwise contained in this publication.

This publication is designed to provide accurate and authoritative information with regard to the subject matter covered herein. It is sold with the clear understanding that the Publisher is not engaged in rendering legal or any other professional services. If legal or any other expert assistance is required, the services of a competent person should be sought. FROM A DECLARATION OF PARTICIPANTS JOINTLY ADOPTED BY A COMMITTEE OF THE AMERICAN BAR ASSOCIATION AND A COMMITTEE OF PUBLISHERS.

Additional color graphics may be available in the e-book version of this book.

Library of Congress Cataloging-in-Publication Data

ISBN: 978-1-53617-536-3

Published by Nova Science Publishers, Inc. † New York

CONTENTS

Preface		vii
Chapter 1	Organic Phase Change Materials: Synthesis, Processing and Applications *Swati Sundararajan and Asit Baran Samui*	1
Chapter 2	Compact Reconfigurable Optical Devices Using Phase-Change Materials *Yin Huang, Lanyan Wang, Yuecheng Shen, Changjun Min and Georgios Veronis*	89
Chapter 3	A Review on the Thermal Behavior of Phase Change Materials Integrated in Building Roofs *I. Hernández-Pérez and J. Triano-Juárez*	137
Index		169

PREFACE

In this compilation, after considering solid-liquid transition, techniques required to obtain phase change materials are discussed. Various material combinations based on chemical and physical methods are also discussed, which are adopted to form solid-solid phase change.

Following this, a non-parity-time-symmetric three-layer structure is introduced, consisting of a gain medium layer sandwiched between two phase-change medium layers for switching the direction of reflectionless light propagation.

The concluding chapter discusses the effectiveness of phase change materials in building roofs for the reduction of energy consumption and the improvement of indoor comfort conditions.

Chapter 1 - In the last two decades, thermal energy storage (TES) has gained wide popularity to bridge the gap between energy demand and supply. Phase change materials (PCM) or latent heat storage materials are the most efficient amongst the TES methods due to high storage density within a narrow temperature range. The simple phase change observed is solid to liquid phase such as paraffin wax, which absorbs thermal energy and continues melting at constant temperature till it completely melts. The organic esters, polyethylene glycols, inorganic salts, eutectic mixture etc. are the prominent PCMs. Organic PCMs and particularly, poly(ethylene glycol) (PEG) can be extremely resourceful to prepare wide array of polymers for thermal energy storage. By preparing different polymers, plethora of working temperatures can be achieved to work at user's comfort. After considering solid-liquid transition, the techniques required to contain the PCMs have been discussed. Various material combinations based on chemical and physical methods have been discussed, which are adopted to form solid-solid phase change. Both solid-solid phase change and thermal conductivity are discussed as the application depends on fast exchange of heat. Normally, inorganic blend components provide higher thermal conductivity as compared to organic counterparts. Both have been incorporated for discussion. Chemical reaction of functional groups on PCM chemical structure can be done with a large number of reactive molecules to form polymeric PCMs such as, polyester, epoxy, polyurethane etc. This has been incorporated with extended discussion. Finally, specific application possibilities have been explored and discussed.

Chapter 2 – The authors introduce a non-parity-time-symmetric three-layer structure, consisting of a gain medium layer sandwiched between two phase-change medium layers for switching of the direction of reflectionless light propagation. The authors show that for this structure unidirectional reflectionlessness in the forward direction can be switched to unidirectional reflectionlessness in the backward direction at the optical communication wavelength by switching the phase-change material $Ge_2Sb_2Te_5$ (GST) from its amorphous to its crystalline phase. The authors also show that it is the existence of exceptional points for this structure with GST in both its amorphous and crystalline phases which leads to unidirectional reflectionless propagation in the forward direction for GST in its amorphous phase, and in the backward direction for GST in its crystalline phase. The authors further switch photonic nanostructures between cloaking and superscattering regimes using phase-change materials at mid-infrared wavelengths. More specifically, the authors investigate the scattering properties of subwavelength three-layer cylindrical structures in which the material in the outer shell is the phase-change material GST. The authors first show that, when GST is switched between its amorphous and crystalline phases, properly designed electrically small structures can switch between resonant scattering and cloaking invisibility regimes. The contrast ratio between the scattering cross sections of the cloaking invisibility and resonant scattering regimes reaches almost unity. The authors then also show that larger, moderately small cylindrical structures can be designed to switch between superscattering and cloaking invisibility regimes, when GST is switched between its crystalline and amorphous phases. The contrast ratio between the scattering cross sections of cloaking invisibility and superscattering regimes can be as high as ~93%.

Chapter 3 - The materials of the building envelope significantly influence heat gains and heat losses. The thermal protection offered by building materials consists of insulating, reflecting, or storing heat. Based on heat storage, phase change materials (PCM) use the heat gain for their phase change, which requires large amounts of energy per unit of volume. This feature allows increasing the heat storage capacity of the building envelope in the order of tens of times. For specific weather conditions, the thermal efficiency of a PCM when integrated in the building envelope depends on the phase change temperature, the latent heat, the volume of material, the position within the building envelope, and the orientation of the PCM, as well as the thermophysical properties of the other elements of the building envelope. Because building roofs have prolonged exposure to solar radiation, they are one of the primary sources of energy gains, especially in regions with warm weather. Thus, this chapter aims to review the application of PCMs in building roofs. The design parameters considered when analyzing building roofs with PCM are mentioned as well as which parameters have shown the more significant influence on the thermal behavior of roofs. Several roof-PCM configurations reported in the literature are illustrated. Further, this chapter discusses how effective are those configurations to reduce energy consumption and to improve indoor comfort conditions in buildings.

In: Phase Change Materials
Editor: Ismaël van der Winden

ISBN: 978-1-53617-536-3
© 2020 Nova Science Publishers, Inc.

Chapter 1

ORGANIC PHASE CHANGE MATERIALS: SYNTHESIS, PROCESSING AND APPLICATIONS

*Swati Sundararajan[1] and Asit Baran Samui[2],**

[1]Zuckerberg Institute of Water Research, Jacob Blaustein Institutes for Desert Research, Ben-Gurion University of the Negev, Sede-Boqer, Israel
[2]Department of Polymer and Surface Engineering, Institute of Chemical Technology, Mumbai, India.

ABSTRACT

In the last two decades, thermal energy storage (TES) has gained wide popularity to bridge the gap between energy demand and supply. Phase change materials (PCM) or latent heat storage materials are the most efficient amongst the TES methods due to high storage density within a narrow temperature range. The simple phase change observed is solid to liquid phase such as paraffin wax, which absorbs thermal energy and continues melting at constant temperature till it completely melts.

---

* Corresponding Author's E-mail: absamui@gmail.com.

The organic esters, polyethylene glycols, inorganic salts, eutectic mixture etc. are the prominent PCMs. Organic PCMs and particularly, poly(ethylene glycol) (PEG) can be extremely resourceful to prepare wide array of polymers for thermal energy storage. By preparing different polymers, plethora of working temperatures can be achieved to work at user's comfort. After considering solid-liquid transition, the techniques required to contain the PCMs have been discussed. Various material combinations based on chemical and physical methods have been discussed, which are adopted to form solid-solid phase change. Both solid-solid phase change and thermal conductivity are discussed as the application depends on fast exchange of heat. Normally, inorganic blend components provide higher thermal conductivity as compared to organic counterparts. Both have been incorporated for discussion. Chemical reaction of functional groups on PCM chemical structure can be done with a large number of reactive molecules to form polymeric PCMs such as, polyester, epoxy, polyurethane etc. This has been incorporated with extended discussion. Finally, specific application possibilities have been explored and discussed.

1. INTRODUCTION

The world is facing a huge energy crisis due to the depletion of fossil fuels, rapidly expanding economies and growing populations. The research on renewable sources of energy has increased tremendously and one of the ways to meet these demands is by using efficient and economical thermal energy storage systems. Thermal energy storage (TES) can be achieved by sensible energy storage, latent energy storage and thermochemical energy. Among these methods, latent energy storage using phase change materials (PCM) are the most efficient energy systems due to high storage density under isothermal conditions [1, 2]. PCMs absorb heat from the surroundings during the melting process and the heat is released back to surroundings during the crystallization process as shown in Figure 1.

As the material is heated, PCM absorbs heat with the collapse of the solid structure and undergoes a phase transformation from solid to liquid phase. While cooling, this energy is released back into the surroundings during the crystallization process making the system reversible and maintenance free. The crystallinity of PCM is favoured by regular and symmetrical structure with fewer short branches and high degree of stereoregularity [3]. An ideal PCM should have the characteristics of high transition enthalpy, suitable transition temperature, cycling stability, good thermal conductivity, low cost, non-toxic nature, and chemical and thermal stability. The heat storage capacity of a PCM can be determined by using the Equation 1 [4].

$$Q = \int_{T_i}^{T_m} mC_p dT + ma_m \Delta H_m + \int_{T_m}^{T_f} mC_p dT \qquad (1)$$

PCMs can be of three types depending on the nature of material: organic, inorganic and eutectic PCMs as presented in Figure 2.

Inorganic PCMs offer high thermal storage capacity with good thermal conductivity but can undergo phase segregation with repeating cycles, subcooling and can cause corrosion to the metallic containers. Organic PCMs have relatively lower phase change enthalpy and thermal conductivity. The advantages with organic PCMs are that they have little or no subcooling issues, thermal stability, non-corrosive nature and chemical stability. Inorganic PCMs include the hydrated salts largely whereas the organic PCMs include paraffins, fatty acids, polyhydric alcohols and sugar alcohols. Eutectics can be of the inorganic or organic type using mixtures of two or more PCMs. Though these eutectics are stable from phase separation and offer good phase change enthalpy but extensive research is required to find eutectics with phase stability and suitable transition temperatures [5, 6]. Some of the most commonly used PCMs have been listed in Table 1. The applications of these PCMs can be extended towards thermoregulating textiles, active and passive solar energy storage, thermal comfort in buildings, temperature-controlled greenhouses, solar cooker, waste heat recovery, temperature sensitive transport of biopharmaceutical products and therapeutic packs. More recently, many novel applications have emerged in the field of nanomedicine, anti-icing coatings, barcoding and biometric identification [1, 7-9, 10].

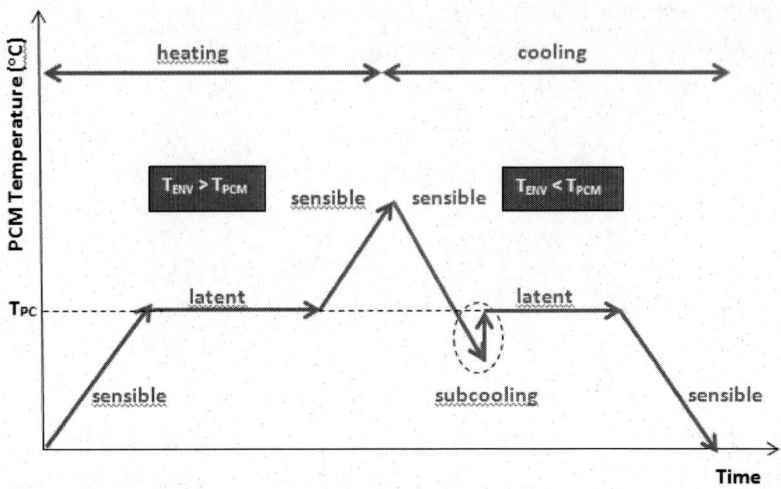

Figure 1. Schematic representation of working of PCM.

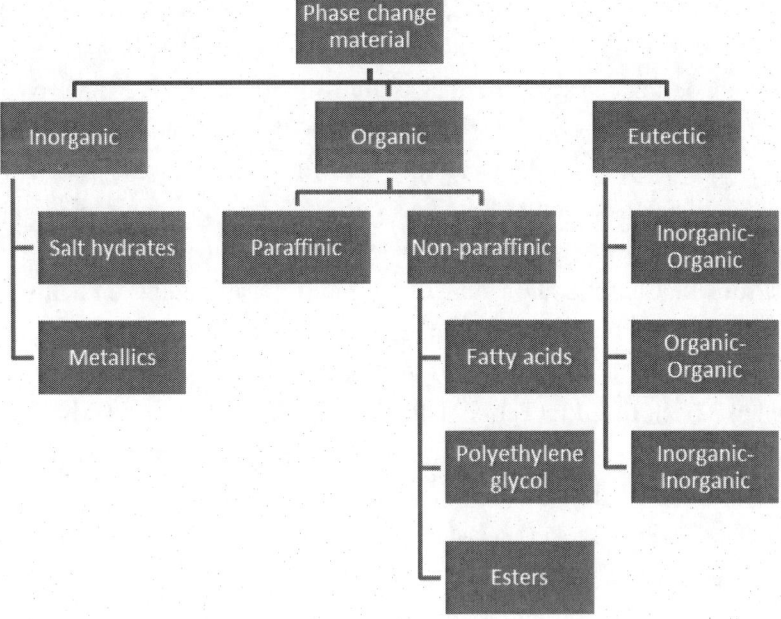

Figure 2. Classification of PCMs.

Phase change transition can be of solid-gas, liquid-gas, solid-liquid and solid-solid categories. Solid-gas and liquid-gas transitions are not feasible for use in TES due to the large volume change involved in the transition to gas phase and the difficulty in its containment. Solid-liquid PCMs undergo limited volume change and have good thermal energy storage properties. Solid-liquid phase transitions are effective TES systems but pose the disadvantages of liquid phase containment and poor thermal behaviour with cycling. The solid-solid PCMs (SSPCMs) have low thermal energy storage capacity compared to solid-liquid PCMs but can be used directly without any need for containment, so that no liquid leakage occurs [11]. The SSPCMs can be prepared by using both physical and chemical methods. The physical methods include blending and impregnation, adsorbing and soaking whereas the chemical methods include grafting, copolymerization and crosslinking reactions with the solid-liquid PCM. The hard polymer segment acts as a supporting matrix and prevents the leakage of the liquid phase. This chapter is divided into three sections with first section comprising of use of paraffin as PCM and second section comprising of PEG, fatty acids and esters as PCM. The third section focusses on the applications of PCMs.

Table 1. List of commonly used PCMs.

Inorganic PCMs	Organic PCMs	Eutectics
$CaCl_2.6H_2O$	Paraffin C_{14}	Capric + lauric acid (65+ 35 mol%)
$Na_2CO_3.10H_2O$	Paraffin C_{16}-C_{18}	Mystiric + Capric acid (34+ 66 mol%)
$Mn(NO_3)_2.6H_2O$	Capric acid	Capric + lauric acid (45+ 55 mol%)
$LiClO_3.3H_2O$	Stearic acid	Urea (37.5%) + acetamide (63.5%)
$Na_2HPO_4.12H_2O$	Palmitic acid	$MgNO_3.6H_2O$ (58.7%) + $MgCl_2.6H_2O$ (41.3%)
$KF.4H_2O$	Oleic acid	$CaCl_2.6H_2O$ (66.6%) + $MgCl_2.6H_2O$ (33.3%)
H_2O	Polyethylene glycol 200-10000	$CaCl_2$ (48%) + NaCl (4.3%) + KCl (0.4%) + H_2O (47.3%)
$CaBr_2.6H_2O$	Erythritol	$Ca(NO_3)_2.4H_2O$ (47%) + $Mg(NO_3)_2.6H_2O$ (33%)
NaCl	1-tetradecanol	
KNO_3	Naphthalene	

2. Paraffin as Phase Change Material

Paraffin waxes, as solid-liquid PCMs, are widely applied because of their interesting properties such as, small amount of supercooling, nontoxicity, non-corrosiveness, chemical stability, and high latent heat energy [12]. However, the practical applications are not so encouraging due to their inherent drawbacks including low thermal conductivity and leakage during the melting process. By preventing the leakage during phase change and increasing the heat-transfer ability is prerequisite for possible application of paraffin waxes. Various methods for containment have been explored, such as impregnation of PCMs into a polymer matrix [13-15], use of a porous or layered material [16, 17], and microencapsulated PCMs. Phase-change microcapsules (microPCMs) have been developed and used. MicroPCMs, with typical core-shell structure, have PCMs as the core surrounded by organic or inorganic material as the shell. MicroPCMs have been used in solar energy storage [18, 19], air conditioning [20], building energy conservation [21], and in thermal-regulating fibres and textiles [22].

2.1. Microencapsulation of Paraffin Wax

Microencapsulation is an effective way to use PCMs as core, which is covered with shells made from polymer materials. The resulting material called PCM microcapsule can prevent the leakage of interior PCMs and increase the heat transfer area [23]. Various strategies are followed to strengthen the shell, to increase the thermal conductivity and add some other attributes to the microcapsule.

The microcapsules with silver nanoparticles can be fabricated by using in situ polymerization, with aminoplast as the wall and bromo-hexadecane as PCM core [24]. The distribution of silver nanoparticles on the surface increases shell toughness, mechanical strength and thermal stability of the microcapsules.

Microencapsulated phase change materials using melamine–formaldehyde resin/SiO$_2$ as shell are made by using 1-hexadecanol as PCM core [25]. SiO$_2$ particles, organically modified by using dimethyldichlorosilane, are employed to stabilize the pickering emulsion, and in situ polymerization of melamine and formaldehyde is carried out to form hybrid shell. The employment of SiO$_2$ results in a higher utility of shell material and its impregnation on shell wall strengthens the shell, enhances the thermal reliability and conductivity, and improves the penetration resistance.

MicroPCMs based on an n-octadecane (C$_{18}$) core and melamine-urea-formaldehyde (MUF)/diatomite hybrid shell can be synthesized by using in situ polymerization [26]. The mechanical properties of the microPCMs are improved largely by the addition of a moderate amount of diatomite. However, the enthalpy and encapsulated efficiency (η) decrease slightly. Water-absorbing polymer shell of poly(2-hydroxy methacrylate) (poly(HEMA)) is used to coat capsule of amino functionalized silica particle containing paraffin (Figure 3) to endure the volume change with repeated cycling, ensure high sealing tightness and good flexibility [27]. The microPCM with 455 μm diameter has enhanced thermal conductivity of 0.47 W m^{-1} K^{-1} compared to dry poly(HEMA). Further, the PCM has high encapsulation efficiency and ratio of 97.7% and 97.3% with latent heat of 167 J g^{-1}.

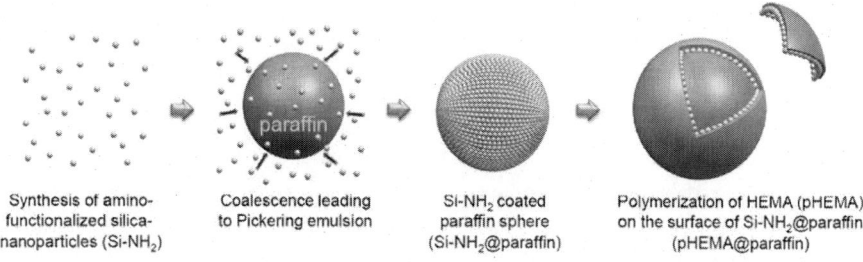

Synthesis of amino-functionalized silica-nanoparticles (Si-NH$_2$)

Coalescence leading to Pickering emulsion

Si-NH$_2$ coated paraffin sphere (Si-NH$_2$@paraffin)

Polymerization of HEMA (pHEMA) on the surface of Si-NH$_2$@paraffin (pHEMA@paraffin)

Figure 3. Schematic representation of preparation of paraffin encapsulated poly(HEMA). (Reprinted (adapted) with permission from American Chemical Society; Do et al. 2015; copyright (2015) American Chemical Society).

The micro PCMs with inorganic shells provide excellent mechanical strength. However, excess rigidity leads to poor endurance during practical applications. The problem can be addressed by using an organic-inorganic hybrid shell as it can be designed to have a synergetic combination of the mechanical properties of inorganic shells and the flexibility of organic shells. The mechanical properties of the microPCMs can be improved by the addition of a moderate amount of modified carbon nanotubes (CNTs) into the polymer system. Thus, the microPCMs based on an n-octadecane (C18) core and a melamine–urea–formaldehyde (MUF) shell, along with O_2-plasma-modified multi-walled carbon nanotubes (MWCNTs), exhibit remarkable thermal and mechanical properties [28]. Of course, there are slight decreases in enthalpy and encapsulated efficiency.

2.2. Paraffin Wax Composites and Nanocomposites

Lightweight polymer composites combining high mechanical properties and TES capability are the materials for applications in different areas, such as the automotive industry, where the diffusion of lightweight structures can disturb the thermal management of the environment in the cockpit, or the portable electronics field, where the space constraint does not permit setting up of cooling system. Structural laminates can be made by combining an epoxy resin, paraffin as PCM, which is stabilized with CNTs, and reinforcing carbon fibers [14]. The stabilized paraffin maintains inherent ability to melt and crystallize in the laminates even after repeated thermal cycling. The melting enthalpy of the composites is proportional to the paraffin weight fraction reaching a maximum value of 47.4 J cm^3. Moreover, the thermal conductivity of the laminates through thickness direction increases with the content of CNT-stabilized PCM. The dynamic mechanical analysis exhibits sharp drop of storage modulus at T_g as well as at T_m of paraffin. Similarly, there are two tanδ peaks corresponding to similar transition as in the case of storage modulus. Paraffin containing samples show a progressive failure during application of load in tensile test. A sequence of drops and plateaus in the load-displacement curves indicates mechanical energy absorption capability of the laminates.

TES composites can be designed with high-performance epoxy matrix together with paraffinic PCM and the CNTs responsible for confining paraffin [15]. For confining paraffin in CNTs, the latter is added to molten paraffin under mechanical stirring at 500 rpm for 30 min and cooled to ambient temperature by pouring in to silicon moulds. For shape stabilization, a minimum of 10 wt% CNT is required for containing paraffin. The CNT-confined paraffin, thus prepared is added to epoxy resin and the mix is cured at ambient temperature followed by post curing at high temperature. Blending process does not affect the inherent melting and crystallization peak temperatures. However, without CNTs, the paraffin exudes out as the confinement action is missing in the blend. Paraffin loading in epoxy lowers the elastic modulus and flexural strength of the epoxy matrix with a trend in accordance with the mixture rule. The electrical resistivity decreases with loading of CNT.

2.3. Wax in Porous Carbon Materials

2.3.1. CNT Sponge

Carbon nanotube sponge (CNTS) as a porous scaffold is also used to encapsulate paraffin wax (PW) to make an electrically conductive composite with enhanced phase change enthalpy and thermal conductivity [29]. Efficient thermal energy storage in this composite can be realized via electro heat conversion or light absorption. The sponges can recover its original shape without visible plastic deformation even after large compressive strain or deformation strain for any arbitrary shape. Instead of molten wax, wax solution is dripped on sponge for infiltration in to pores due to high viscosity of former as illustrated by Figure 4 (a). The porous CNTS has pores of the order of few hundreds of nm to 1 μm and this internanotube spacing is filled by PW as shown in Figure 4 (c) and (d). The composite is so stable that by keeping it above melting temperature for two weeks, no liquid separates. Also, the enthalpy values remain slightly lower at PW loading below 87%. However, with increasing PW content, the value becomes higher than pure PW. Six-fold increment of thermal conductivity of PW occurs after making composite. There are plenty of C-H bonds in the PW and delocalized π electrons on the surface of nanotubes, which lead to extensive C-H ...π interactions at the PW-nanotube interfaces. This interaction affects the phase change enthalpy of paraffin. Further, the compressive stress applied by the nanotubes on the melting PW during its volume expansion also inhibits the solid-to-liquid transition resulting in increased system enthalpy.

Vertically aligned CNT arrays are highly porous, and because of high absorption of sunlight it appears black in the original state. Further, the enhanced photon absorption, due to presence of CNT, the PW in the composite can be heated. When current passes through the composite, the resistance of individual nanotubes and also the contact resistance in the network generate Joule heat. The main heat transfer mechanism involves solid-to-solid heat conduction at the PW nanotube interface [30].

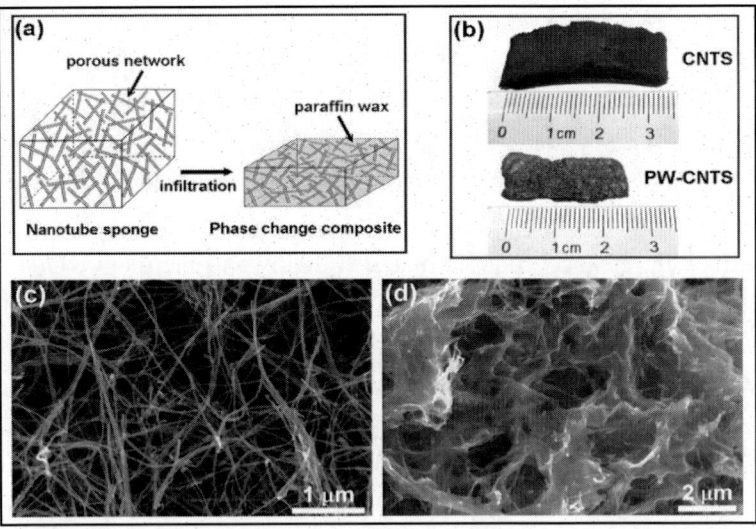

Figure 4. (a) Schematic for preparation of CNTS-PW composite, (b) Photograph of original CNTS and CNTS after PW infiltration, (c) SEM of sponge showing porous structure and (d) SEM image of CNTS-PW composite. "Reprinted (adapted) with permission from ACS Nano 6, 12, 10884-10892. Copyright (2012) American Chemical Society."

The electroheat storage efficiency (η) can be calculated by the ratio of stored heat in PW [total mass of PW enclosed in the composite (m) times its enthalpy (ΔH)] and the received electrical energy during the phase change period [product of voltage (U), current (I), and time (t)]. The energy storage efficiency in the system is defined as η = (m.ΔH)/U.I.t on the basis of assumption that all the encapsulated PW material has participated in the phase change. Sunlight-to-heat storage efficiency of more than 90% is possible by simulating light irradiation at a wavelength (half-wave width of 80 nm) corresponding to maximum absorption on dye-grafted phase change material [31].

2.3.2. Graphene Foam

Graphene foams (GF) with three-dimensional (3D) interconnected network are being used to prepare composite PCMs, which exhibit enhanced thermal resistance for heat transfer from the PCM to the GF walls. A 3D hierarchical GF (HGF) can be obtained by filling the pores of GF with hollow graphene networks, which is then used to prepare a paraffin wax (PW)-based composite PCM [32]. The thermal conductivity of PW/HGF composite PCM is 87% and 744% higher than that of the PW/GF composite PCM and pure PW, respectively. The shape stability, negligible change in the phase-change temperature and high thermal energy storage density (95% of pure PW), coupled with good thermal reliability, and chemical stability characterize the composite. In addition, the composite PCM allows light-driven thermal energy storage with a high light-to-thermal energy conversion and storage efficiency.

2.3.3. Intercalated Wax

Composite PCM, comprising organic montmorillonite (OMMT)/paraffin/grafted MWCNT, can be synthesized by using ultrasonic dispersion and liquid intercalation [33]. Paraffin is intercalated into OMMT interlayer, and the grafted MWCNT is well dispersed in the interlayer of OMMT. The latent heat of the OMMT/paraffin/grafted MWCNT is around 47.7 J g^{-1}, which reduces slightly after extended thermal cycles. The thermal conductivity of the OMMT/paraffin increases by 34% after adding grafted MWCNT.

2.4. Structural Laminates with Paraffin Wax

Structural laminates can be designed by combining paraffin wax, stabilized with CNTs, with an epoxy resin and reinforcing carbon fibers [14]. The stabilized paraffin is able to melt and crystallize in the laminates, and the melting enthalpy of the composites is proportional to the paraffin weight fraction (maximum value: 47.4 J cm^{-3}), which is stable to repeated thermal cycling. Presence of CNT increases the thermal conductivity of the laminates through thickness direction. The flexural modulus is only slightly lower due to the presence of PCM.

3. POLYETHYLENE GLYCOL, FATTY ACID AND ESTERS AS PHASE CHANGE MATERIALS

Fatty acids possess higher enthalpy of fusion as compared to that of paraffin wax. Fatty acids exhibit reproducible melting and freezing behaviour and minimum supercooling. The general formula describing all the fatty acid is given by $CH_3(CH_2)_{2n}$-COOH. However, their major drawback is their excessive cost, which is about 2–2.5 times greater than that of paraffin wax. Also, they are mildly corrosive. PEG, having general formula H-(O-CH_2-CH_2)$_n$-OH, is exhaustively studied as phase change material due to high phase change enthalpy, variable transition temperatures, ease of chemical modification, congruent transition behaviour, chemical stability, thermal stability, non-corrosiveness and biocompatibility [34]. The melting temperature and enthalpy of PEG varies with molecular weight. The PCM can be designed by physically blending or chemically attaching it with various organic and inorganic components.

3.1. Electrospun PCM Fiber

To make PCMs with a different design, nanoscience and technology was employed. The new design, as is called a nanofiber, is an ultrafine fiber of PCM/polymer composite that is form-stable or shape-stable in nature [35]. The nanofibers can be formed via electrospinning by application of very high voltage, which is a simple and convenient technique for generating ultrafine fibers with diameters on both the micro- and nanoscales [36]. In general, paraffin waxes, fatty acids, and their blends, polyethylene glycol (PEG), are commonly studied for developing form-stable PCMs. The other component in a form stable composition is the electrospinable polymer which serves as structure-supporting component. Few of the structure supporting polymers tried can be named as polyvinylpyrrolidone [35], cellulose acetate (CA) [37], polyethylene terephthalate (PET) [38]. All the composite fibers possess good thermal stability and reliability. Single fibers and coaxial fibers are generally possible with electrospinning technique. In the second category, cooling of the jet occurs due to solvent evaporation that allows the inner liquid to quickly solidify, leading to its encapsulation in the nanofibers. Also, core–sheath structured PCMs/polymer composite nanofibers can be successfully fabricated via melt-electrospinning method. Another technique, named as multifluidic compound-jet electrospinning is also used. Electrospinning generate ultrafine fibers from a wide variety of polymers, polymer blends and nanoparticle-impregnated polymers that have huge surface-to-volume ratio and can be used for numerous applications in areas such as healthcare, biotechnology, environmental engineering, and energy storage.

The incorporation of PCM in textile was initially targeted for astronauts' space suits in order to provide higher protection against high thermal fluctuations in outer space [39]. PCMs were normally embedded into textile materials with the development of microencapsulation technologies in 1987. Currently, the PCMs are embedded directly inside fibers. Electrospun phase change fibers or ultrafine fibers are expected to be applied as coat or wrap temperature sensitive products in refrigeration equipment and constitute smart food or medical/pharmaceutical packaging most important parameters to consider. The morphology depends on the weight ratio of phase change material and the polymer used for making fibers. Inferior morphology results beyond a particular weight ratio, which depends on the particular type of polymer and PCM. Hydrogen bonding is easily detected by peak shifting of corresponding groups. The enthalpy of phase change is studied by DSC. The values lower than theoretical are due to dilution effect by the presence of fiber forming polymer and hindrance for crystallization by the hydrogen bond formed among them as well as the restriction of crystallization during quenching of electrospun fibers. Thermal stability of the polymer is decreased due to less thermally stable PCM component. Similarly, the mechanical properties of the polymer deteriorate due to the small molecule PCM acting as plasticizer. Few examples are discussed below:

3.2.2.2. Fatty Acid/Polyethylene Terephthalate Composite Fibers

The lauric acid (LA)/polyethylene terephthalate (PET) composite fibers with mass ratio up to 1:1 are normally smooth and cylindrical in shape that indicates good miscibility of the composite fibers [45]. However, higher mass ratio of LA/PET results in non-uniform fiber that is originated from the aggregation of the excessive low molecular weight fatty acid. As a general rule, the enthalpy values of LA/PET composite fibers are lower than the pure LA powders. Of course, the theoretical values and experimental values are almost same. T_m and T_c are not much affected due to composite making.

A series of diacid dioctadecyl esters (DADOEs) and PET composite fibers (375 nm to more than 1µm diameter) are prepared [46]. Bead-free fibers with smooth surface are obtained up to DADOE content of 50% and beyond that there is appearance of solid particles on the fiber. There is no change of T_m and T_c values after making composite fiber, ΔH_m and ΔH_c of the composite fibers with different DADOE contents are lower than those of pure DADOE. The latent heat values for the composite fibers are lower than theoretical values due to quenching of fibers during electrospinning which hinders crystallization.

3.2.2.3. PEG/Cellulose Acetate Composite Fiber

PEG/cellulose acetate (CA) exhibits uniform and cylindrical morphology up to 50 wt% of PEG [47]. A crosslinked membrane-like structure is formed in the fibrous mats and there is presence of many tubercles in fibers. Carbonyl peaks of CA shifts from 1757 cm^{-1} to 1740 cm^{-1} in PEG/CA fibers due to hydrogen bond formation. There is increase in the degree of crystallization with increasing PEG content as evident from the sharper and stronger diffraction peaks with the increasing PEG mass percent. This is also seen with enthalpy, which increases gradually with increasing PEG content. The lower enthalpy results from dilution effect and hydrogen bonding. The ultimate strength and strain decreases with increase in PEG content.

The problem of the thermo-oxidative degradation of PEG is minimized by crosslinking the fibers to isolate PEG from air. By electrospun PEG/CA fibers are crosslinked by dipping into the TDI solution [48]. The fibers form a dense network membrane by the crosslinking that can be confirmed by water resistance test. In crosslinked state, there is restriction of mobility of the polymer chains that result in decreased crystallization and the enthalpy. Infact, the enthalpy reduces by 50% as compared to uncrosslinked composite. The thermal stability increases due to crosslink formation.

3.2.2.4 PEG/Nylon-6 Composite Fibers

Nylon-6/PEG-blended nanofibers exhibit a good compatibility between Nylon-6 and PEG [49]. The addition of PEG increases the viscosity of the spinning solution, which results in excellent morphology with regular shape, uniform distribution and no breakage. The increase of PEG beyond 30% the viscosity decreases that leads to inferior morphology. The characteristic absorption peaks of amide in PA6/PEG-blended nanofibers redshift due to the formation of hydrogen bonding between the N-H or C-O group in PA6 and the C-O or O-H in PEG. Both the phase-transition temperature and enthalpy value of PEG/Nylon-6 nanofibers are lower than those of the PEG powder because of the relatively high melting point of Nylon-6, which remain solid impurity during phase change. The enthalpy decreases gradually with increasing Nylon-6 and thermal stability of the composite remains lower than pure Nylon-6. Tensile strength decreases with increasing PEG while elongation increases.

3.2.2.5. Emulsion-Electrospinning N-Octadecane/Silk Composite Fiber

The ultrafine n-octadecane/silk form-stable phase change composite biodegradable and biocompatible fibers are made by the emulsion-electrospinning method [50]. The emulsion of silk, dissolved in aqueous solution, and n-octadecane is made by using emulsifier. The morphology of fiber changes gradually from the spindle-like bead structure to the continuous fiber structure by varying the silk solution concentration from 19% to 25%. When n-octadecane content in the O/W emulsion reaches 16%, the fibers are not formed due to the easy blocking of the spinneret and the unstable jet flow resulting from the high emulsion viscosity. Because of restriction of n-octadecane chain mobility there is decrease of Tm, Tc and enthalpy of melting (40–60 J g^{-1}), while supercooling extent increases.

3.2.2.6. Multicomponent Composite PCM Fiber

The stability and potential application of solid-liquid PCMs can be improved by incorporating inorganic compounds in the PCM composite matrix. This helps in preventing the molten PCM to flow and also enhances the thermal conductivity of the matrix. The multicomponent matrix can be used for making nano/micro fibers via electrospinning.

3.2.2.7. Polyacrylonitrile (PAN)/Lauric-Stearic Acid (LA-SA)/TiO_2 Composite Nanofibers

The LA-SA eutectics and TiO_2 (0.5%) mixed with PAN solution and electrospun to get composite phase change nanofibers [51]. Up to LA-SA to PAN mass ratio is 0.5:1 the composite nanofibers are quite uniform and fine in diameters. At higher mass ratio the higher viscosity results in coarser fibers and also beads start appearing. The latent heat and melting temperature of composite nanofibers remain much lower than the pure LA-SA eutectic, which is primarily due to hindrance by TiO_2. The thermal stability of the nanofiber increases with increase of LA-SA eutectic due to increase of hydrogen bonding ability to delay the degradation.

3.2.2.8. Flexible SiO_2 Nanofibrous Mats/PCM Composite

Flexible SiO_2 nanofibrous mats can be fabricated as supporting materials for making PCM composite by electrospinning of PVA/SiO_2 composite followed by its pyrolysis to remove PVA. Five types of quaternary fatty acid eutectics as PCMs are incorporated into flexible SiO_2 nanofibrous mats for thermal energy storage/retrieval [52]. The nanofibrous mats with 3D porous structure can absorb a large amount of quaternary fatty acid eutectics with a minimum of 85.1%. The calculated and experimental enthalpies of melting are almost identical for all the compositions.

3.2.2.9. Coaxial Electrospinning

Coaxial electrospinning technique has the advantage of isolating from the atmospheric exposure and damage, while maintaining high enthalpy as observed with microcapsule. Thus, the coaxial electrospinning can offer a straightforward way to encapsulate the PCM in the core and maintain it inside the polymer sheath layer to eliminate the leakage issue. The composite fibers are created by the wet–wet/melt/co-axial electrospinning method. The flow rates of individual components (core and sheath) are very important to get right kind of fiber having minimum defect.

3.2.2.10. Polyacrylonitrile/PCM Coaxial Nano-Fibers

Wearable PAN/PCM sheath/core nano-fibers is made by using coaxial electro-spinning technology [53]. Isopropyl palmitate (IPP) and paraffin oil are used as PCM. The flow rates of 2 mL h^{-1} and 0.1 mL h^{-1} for PAN solution and IPP component, respectively produces the best fiber morphology. At lower flow rate of PAN beads are formed. The DSC analysis indicates that the encapsulation ratio of paraffin oil and IPP inside the nano-fibers are 61% and 28%, respectively.

3.2.2.11. Coconut Oil-Cellulose Beaded Microfibers by Coaxial Electrospinning

Electrospun co-axial microfiber isolates coconut oil in the core surrounded by a cellulose shell [54]. The cellulose-room temperature ionic liquid (RTIL) (1-ethyl-3-methylimidazolium acetate; [EMIM][Ac]) electrospun fiber jet forms an intermediate hydrogel upon entering the coagulation bath by releasing the RTIL. By washing several times with distilled water and water/ethanol mixture the RTIL is removed via swelling of fiber. Finally, the fibers are freeze-dried to obtain dry nonwoven microfiber. Freeze-drying is employed to obtain a structure similar to cotton balls [55]. About 76±10% of the coconut oil is encapsulated in the composite fibers. The composite fibers have a beaded structure with alternating cylindrical and biconical shaped regions having an average diameter of 3000±1000 nm excluding the latter regions. The thermal stability of the composite fiber is lower than the coconut oil. Marginally lower melting temperature of coconut oil in the composite fiber is at (22±0.1 °C), and the crystallization temperatures are at (14.0±1.0 °C) and (8.0±0.1 °C) as compared to neat coconut oil. The coconut oil confined to a smaller axial micropore, the specific heat capacity of solid and liquid oil increases by 98 and 24%, respectively, as compared to the unconfined oil.

3.2.2.12. Dual-Functional Ultrafine Composite Fibers

With the advancement of technology there is always need for one material to have multifunctional properties. PCM materials are needed to take up the heat and melt and later release the heat and solidify. This being purely physical and reversible change, another important property can be added to it which functions under different trigger.

3.2.2.13. Fibers with Dual Function of Phase Change and Luminescence

The thermal stability of chromic and luminescent materials being poor it is difficult to fabricate such fibers through conventional melt-spinning methods. The use of other chromic materials poses another problem. Electrospinning technique is already used to make fibers. A dual-functional ultrafine composite fiber with phase-change energy storage and luminescence properties is successfully prepared by using a parallel electrospinning technique. A thermoplastic polyurethane solid-solid phase-change material (PUPCM) is used to provide temperature stability, while a mixture of polymethyl methacrylate (PMMA) and an organic lanthanide complex (TbL) act as the luminescent component in this dual-functional ultrafine fiber [56]. The ultrafine fibers exhibit good fluorescence properties and emit a bright green light due to the presence of the polymeric PCM.

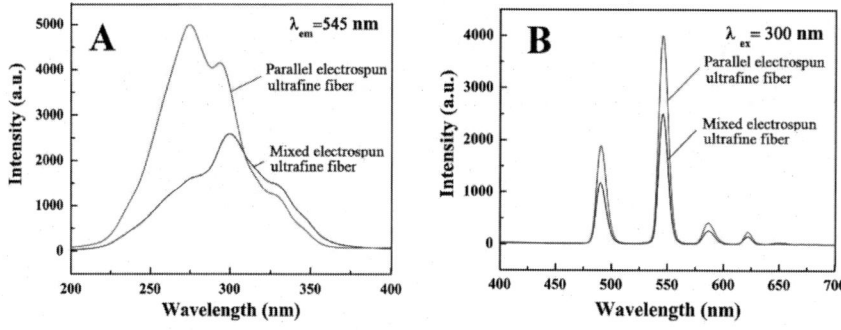

Figure 6. Fluorescence excitation (A) and emission spectra (B) of the parallel electrospun ultrafine fibers.

Figures 6A and B present the the fluorescence excitation and emission spectra of the parallel electrospun ultrafine fibers. The emission spectra of the parallel and mixed electrospun ultrafine fibers are similar. The strongest emission peak intensity of the parallel electrospun ultrafine fibers is 1.6 times higher than that of the mixed electrospun ultrafine fibers. Also, the fluorescence intensity of these new dual-functional ultrafine fibers is higher (2.5 times) than that of PMMA ultrafine luminescent fibers. SEM images of the as-prepared parallel electrospun ultrafine fibers are shown in Figure 7A–D. TEM shown in figure 5E indicates that the of the as-prepared fibers are composed of two parts; the light regions (PUPCM), and the dark parts (PMMA and TbL particles). The phase-change enthalpy of the PCMs is also improved, and the phase-change temperature range decreases from 18–55 °C to 25–48 °C due to the nano-size effects of the ultrafine fibers. As the PMMA mass ratio is increased the fiber surfaces become smoother, and the average fiber diameter decreases. As usual there is decrease of enthalpy of melting for the fiber as compared to pure PCM.

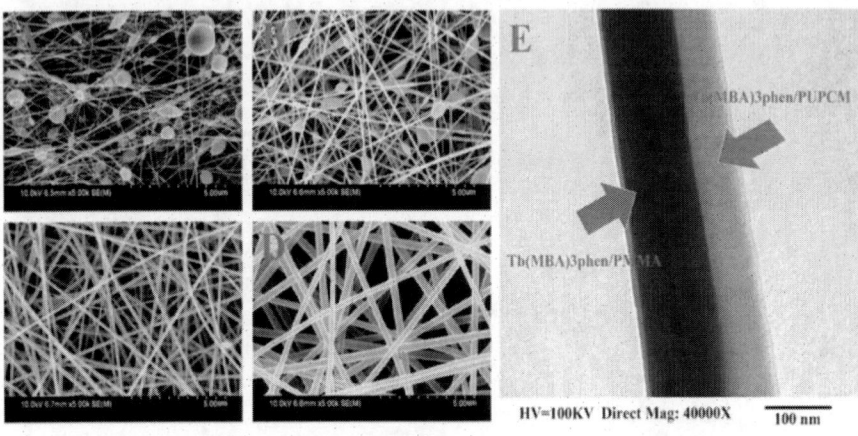

Figure 7. SEM images of parallel electrospun ultrafine fibers with polymer concentrations of 16% (A), 20% (B), 24% (C), and 28% (D); TEM image of fiber (E).

Similarly, PVP/Eu-carboxyl modified PEG complex fibers exhibit good phase-change properties and excellent luminescent properties [57].

3.2.2.14. Thermochromic Phase Change Nanofibers

PCMs in various forms, such as microcapsules and others can be used for making thermal management of textiles. Making nanofiber gives the best possible form for incorporating in textile. In addition, thermochromic material can be incorporated in the fiber to impart temperature sensing property as the fiber will change color at specific temperature. A thermochromic phase change nanofibers /woven composite material is fabricated with a sandwich structure having the first layer porous weaves, the second layer thermochromic phase change nanofibers, and the last layer cotton weaves [58]. A simple suture & hot-pressing treatment are employed for connecting these three layers. Thermochromic powder (TP)/lactic acid (LA)/PET fibers exhibit roughness due to presence of microcapsule TP nanoparticles which do not dissolve in the solution. The fiber shows definite change of color as a function of temperature and also act as phase change material.

3.2.2.15. Fiber with Paraffin Wax as Core Layer and Polyacrylonitrile as Sheath Layer

Multifunctionality can be imparted by incorporating PCM as well as material which can absorb solar radiation. The clothes are made by using smart textile with organic PCMs PW as core layer and PAN as sheath together with hexagonal $Cs_{0.32}WO_3$ nanoparticles as a heat absorbent to utilize higher amount of solar energy in near infrared region (NIR) together with functioning as heat storage medium [59]. The polyacrylonitrile (PAN) and the PW solution were mounted in the double spinneret to produce the sheath and core layer, respectively. The fibers surface is smooth and no PW can be seen that indicates effective encapsulation of PW. Both melting and crystallization enthalpy decreases rapidly while transition temperatures are not affected much. It also exhibits excellent heating/cooling cycling stability beyond 500 cycles.

3.2.2.16. Nanofibrous PCM Mat

Nanofiber mat can be prepared with PCM material by creating hole on the fiber after electrospinning it. In this method two polymers viz., cellulose acetate and polyvinylpyrrolidone (CA/PVP) are electrospun to bicomponent nanofibers and PVP is then selectively removed via dissolution by water treatment to leave pores in the fiber that enhances PCM component incorporation ability [60]. Capric–myristic–stearic acid (CMS) (77: 18.5; 4.5 by wt.) ternary eutectic mixture is absorbed (up tp 83%) in the mat by dipping it under molten condition. The nanofibers are perfectly cylindrical and quite smooth with CMS covering the pores. A marginal decrease of transition temperature and about 20% decrease in enthalpy are observed.

Nanofibrous PCM mats can be prepared by directly incorporating the PCM by absorbing it in to the fiber. Polyvinyl butyral–Poly (acrylic acid) (PVB-PAA) can be prepared via electrospinning process and then different kinds of fatty alcohols are absorbed into electrospun mats [61]. By incorporating in nanofiber there is slight reduction of melting and freezing temperature of fatty alcohols. Enthalpy values for melting and freezing are also slightly lower (223-241 J g^{-1} and 215-239 J g^{-1}, respectively). The fibers are uniform, homogenous, bead free with smooth surface.

3.3. Applications of Electrospun Fiber

Maintaining the cold chain ensures food safety and food quality. However, there can be temperature variation, which is likely to spoil food and affect the food quality. The PS multilayer-based heat storage structures are based on PS films coated with polycaprolactone (PCL)/PCM electrospun layers [62]. Heat management materials comprising PS trays coated with the electrospun PS/PCM layers are developed [63]. The PS/PCM coating deposited on the PS trays is annealed at 145 °C for 1.5 min using a hot-plate hydraulic press to enhance the interlayer adhesion. The surface images, showed a dense but opened structure with many beaded areas within the fibrous mat, and the cross-section images confirms compactness. As usual, the enthalpy of melting and enthalpy of crystallization decreases with increasing PS in PS/PCM fiber.

3.4. PCM Electrospun Fibers for Promoting Neurite Outgrowth under Controlled Release

It is necessary to control and manage the release of growth factors from the scaffolding materials under both in vitro and in vivo conditions. This is easily possible by integrating electrospun fibers with a controlled release system based upon a stimuli-responsible material. The PCM (mixture of lauric and stearic acid at a mass ratio of 8:2) dissolved in a mixture of ethanol and dichloromethane (20:80 by vol.) at a concentration of 20% is used as the outer solution. The payload was dissolved in 0.5 wt% aqueous gelatin solution and used as the inner solution. Both are electrospun in co-axial mode [64].

The reversible phase change nature of PCM allows the payload release in a pulsatile mode through multiple on/off heating cycles. Also, by sandwiching the PCM microparticles (coloaded with nerve growth factor (NGF) and a NIR dye) between two layers of electrospun fibers, the NGF release on-demand is made possible by triggering with photothermal heating.

3.5. Form-Stable PCMs

Form-stable composites can be developed by physical blending of soft segment with a higher melting polymeric substrate such that the soft segment is embedded in polymer matrix preventing its leakage during the phase transition from solid to liquid phase, as long as the temperature is maintained below melting point of polymer matrix.

3.6. Organic Form-Stable PCMs

3.6.1. Form Stable Composite with Ethylene Homo/Co-Polymers

Various amounts of a paraffinic wax can be dispersed by melt mixing in an ethylene/propylene diene monomer (EPDM) rubber matrix [13]. The compounds are then vulcanized to obtain shape-stabilized rubber materials for thermal energy storage. There is reduced rate of vulcanization of the EPDM matrix due to presence of rubber. Homogenous distribution of the wax particles within the rubber allows the melting enthalpy to remain almost unchanged and remain so even at higher concentrations and after repeated thermal cycles. The incorporation of polyethylene wax had a positive effect (increasing proportionally to its content) on the mechanical properties of the EPDM matrix, as documented from both the dynamical and the quasi-static tensile tests.

Phase change materials based on soft paraffin wax and polyethylene blend is made by mixing copper powder with blending and then melt pressing techniques [65]. The melting peaks at lower wax loading are clearly defined as lower temperature shoulder and higher temperature peak. However, at higher wax loading, the peaks are exactly opposite. Shorter LDPE chains and/or branches co-crystallize with the wax, making this king of melting. This effect becomes less pronounced with increasing wax content. The melting peak temperature of LLDPE in the blends decreases with wax loading, which is due to the molten wax acting as a plasticizer in the LLDPE matrix. The crystalline nature of wax enhances the modulus of PE. Presence of Cu micro-particles did not significantly change the thermal behaviour of the blends. By increasing copper content, the thermal stability of the composite is increased.

3.6.2. Composites with Cellulose and Cellulose Derivatives

There are constant efforts to use natural polymers for commercial application so that non-biodegradable polymer pollution is minimized. Cellulose based polymers are abundant in nature. Further, these are polar in nature, thermally and chemically stable, which are very much required for making form-stable PCM blends.

The earliest known blends of PEG were prepared with cellulose. Cellulose extracted from wood pulp is blended with PEG at various proportions [66]. Intermolecular hydrogen bonding between PEG and cellulose prevents the flow of molten PEG that results in solid-solid transition of PEG from crystalline to amorphous state at temperatures above the melting point of PEG. Crystal size of PEG is usually smaller in the blend. This has direct effect on the enthalpy of transition and transition temperature, which are lower at 120 J g^{-1} and 52°C respectively. Another renewable polymer cellulose diacetate (CDA) can be modified with PEG by both physical and chemical bonding methods [67]. By chemical bonding method the PEG molecules are constrained, which has direct effect on its crystallization ability. The solid-solid PCM of CDA and PEG exhibits lower enthalpy of 74 J g^{-1}, which is due to lower crystallinity. On the other hand, the physical blending method gives rise to solid-liquid PCM with much lower constraints on crystallization of PEG. The blend has higher enthalpy of 105 J g^{-1} and a phase transition temperature of 40°C. It is also the PEG content which decides the phase change pattern. Keeping the PEG content below 85%, it is possible to retain solid-solid phase change during melting (155 J g^{-1} at 52 °C) [68]. PEG is blended with cellulose acetate in the presence of microwave (Figure 8 (a)) reducing energy and solvent consumption and enhancing the encapsulation efficiency with 96.5 wt% PEG [69]. The latent enthalpy varies from 23.5-155.4 J g^{-1}. The network structure of the blend is visible at lower concentrations and globular morphology is prominent at higher concentrations of PEG (Figure 8(b)). Also, the presence of cellulose acetate reduces the hydrophilic nature of PEG.

The type of phase transition of polyethylene oxide (PEO) and cellulose polymers blends depends on the type of the latter [70]. Thus, the PEO-cellulose acetate blends show solid-liquid phase transition for all concentrations, whereas solid-solid phase transitions are possible for 25 and 50 wt% PEO in PEO-carboxymethyl cellulose blends, 25 wt% PEO in PEO-cellulose ether blends and for the entire range of concentration of PEO (25-75 wt%) in PEO-cellulose blend. Strong hydrogen bonding interactions between PEO/cellulose and PEO/cellulose derivatives blends make the solid-solid phase transition possible. At 1:1 (w/w) compositions of PEO/cellulose acetate and PEO/carboxymethyl cellulose blends, synergistic transition enthalpy is possible. Blends of PEG with biodegradable natural polymers such as, cellulose, agarose and chitosan, having strong hydrogen bonding interactions, form stable PCMs [71]. Three-dimensional porous cellulose/graphene nanoplatelets (GNPs) aerogels are fabricated in which PEG is vacuum impregnated to prepare phase change composites [19]. The GNPs, having very high thermal conductivity, increases the heat carrying capacity to large extent (at 5.3 wt%, thermal conductivity of 1.35 W m^{-1} K^{-1}). Very high fusion enthalpy of 156 J g^{-1} and corresponding crystallization enthalpy of 149 J g^{-1} can be realized by loading 89.2 wt% PEG in the composite. With large surface to volume ratio and small diameter, the insoluble and fibrous carboxy methyl cellulose-1 fibres can be treated with eutectic mixture of lauric and stearic acid [72]. The light weight composite exhibits latent heat of 115 J g^{-1} with very good thermal reliability.

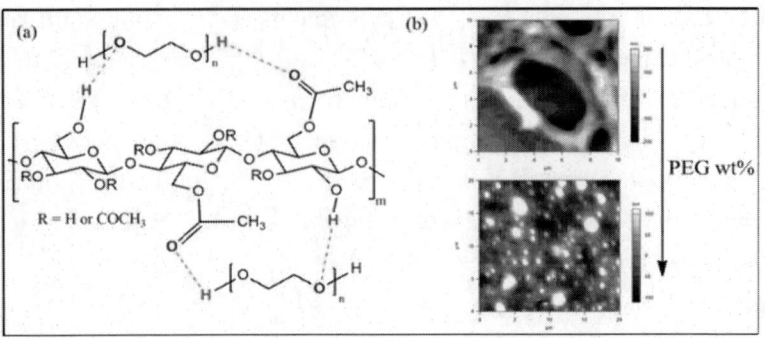

Figure 8. (a) Interaction of PEG and cellulose acetate in composite and (b) AFM images of form-stable PEG-CA blend with increasing concentration of PEG. (Reprinted with permission from Elsevier; Swati et al. 2016).

3.6.3. Acrylate Based Composites

PEG and acrylates being moderately polar in nature, the blends/composites are compatible. The acrylates are thermally stable and do not impart any colour or negative effect. Blends of PEG with acrylic polymers such as, polymethyl methacrylate (PMMA), Eudragit S and Eudragit E can be prepared with enthalpy values ranging from 141-149 J g^{-1} [73]. These blends exhibit solid-solid phase transition till 80 wt% encapsulation of PEG. The blends have good thermal and chemical reliability, as the enthalpy of fusion reduces by only 10% with no chemical degradation even after 3000 thermal cycles. The interactions of acid groups of acrylic acid polymers, such as poly(acrylic acid) (PAA) and poly(ethylene-co-acrylic acid) (EcoA), with PEG lead to the formation of interpolymer complexing blends, stabilized through hydrogen bonds [74]. For the PEG-PAA blend, latent enthalpy can be observed only at high concentrations of PEG (≥75 wt%), whereas in PEG-EcoA blends, the latent enthalpy values are observed at all compositions. With low molecular weight PEG (≤ 1000), the incompatibility in PEG-EcoA blends lead to seepage. A series of binary fatty acid eutectics of capric, lauric, myristic and stearic acid can be developed to prepare form-stable PCMs with PMMA supporting matrix [75]. The combination of lauric and myristic acid with 70 wt% composition within the PMMA matrix is most suitable for use as form-stable PCM with storage enthalpy of 113 J g^{-1}.

Although, the versatility of the PEG-PMMA composite is established but the commercial feasibility depends on the thermal conductivity of the composite. To overcome the low thermal conductivity of the organic composite PCMs, thermal fillers can be used to improve the thermal conductivity. Metal-based thermally conductive filler viz. aluminium nitride is added during in-situ polymerization of methyl methacrylate to prepare a composite withstanding maximum encapsulation of 70 wt% PEG [76]. By varying the mass fractions of aluminium nitride in the composite in the range of 5-30%, the thermal conductivity is increased from 0.253 to 0.389 W m^{-1} K^{-1}. However, increasing mass fractions of aluminium nitride results in reduced latent enthalpy from 116 to 79 J g^{-1}. On the other hand, the volume resistivity increases from 0.26×10^{10} to 5.92×10^{10} Ω cm for PEG/PMMA composite by adding 30 wt% aluminium nitride, which promotes the electric insulation capabilities of the PCM. Therefore, this composite PCM can be used for thermal management of electronic devices that have acceptable electrical insulation properties. Carbon based fillers are gaining importance due to excellent thermal and electrical properties, large surface area, low density and significant chemical stability. Thus, graphite nanoplatelets (GnP) are added as conductive fillers via ultrasonic treatment to form PEG-PMMA form-stable PCM networks that have improved thermal and electrical conductivity [77]. For PEG-PMMA composite, a maximum of 70 wt% PEG can be incorporated. Due to high aspect ratio of GnP, thermally conductive pathway is formed in the composite. As a result, the thermal conductivity of PEG-PMMA/GnP composite increases by about 20 times to 2.339 W m^{-1} K^{-1} from that of pure PEG-PMMA composite with incorporation of 8 wt% GnP. The electrical conductivity of the composite is also improved due to electrically conducting nature of GnP. Percolation threshold is reached with only 2 wt% GnP and the corresponding electrical conductivity is 10^{-4} S cm^{-1}. The addition of GnP aids the crystallization of PEG and therefore, the supercooling extent reduces with increase in thermal stability. This type of composite is considered a potential material for the applications of EMI shielding, anti-electrostatic material and bipolar plates in proton exchange membrane fuel cells.

3.6.4. Polyurethane Blends

Crosslinked polyurethanes (PU) are considered as a container for the PCM with an aim to obstruct the migration of PCM with repeated use and give way to synergistic phase change effect allowing wider transition range. The solid-solid PCM with urethane linkage is prepared by synthesizing the prepolymer from PEG and MDI, which is further reacted with glucose. The reaction mixture is then blended with extra PEG and cured with heat over extended period to get form-stable PCM [78]. The synthesized polymeric PCM exhibits solid-solid phase change and a synergistic effect is observed due to presence of PEG in the crosslinked polymeric chain as well as in blend. An improvement in enthalpy of 21.3% over the polymeric PCM can be achieved. An environmentally friendly, cost-effective method is adopted to prepare polyurethane/paraffin composites by avoiding the use of toxic organic solvents [79]. This method allows maximum encapsulation of a series of paraffin in the hard PU segment made with PEG 10,000. Fusion enthalpy of 141 J g^{-1} and the transition temperatures ranging from 20 to 65°C, exhibited by this combination, is suitable for multiple energy storage applications. In these polyurethane composites, the chain extender is a vital component that is responsible for increasing the phase separation that enhances crystallization of soft segments [80]. However, with PU based graphene composites of SSPCMs without chain extender, higher heat of fusions, higher thermal stability and better thermal reliability are observed [81].

Crosslinked polyurethane prepared by using PEG and branching unit, xylitol can be used as a supporting matrix to prepare lauric acid/PU composite having high melting enthalpy of 125 J g^{-1} and reduced supercooling as a result of overlap of phase transitions of PU and lauric acid [82]. Using the same crosslinked polyurethane system with PEG as supporting material, paraffin can be utilized as the working material [83]. The system exhibits synergistic phase change with enthalpy reaching 210.6 J g^{-1} at an encapsulation percentage of 74 wt%. Paraffin promotes microphase separation by reducing hard-soft segment compatibility. Also, by integrating dye molecules with the polyurethane matrix, direct conversion of visible light to thermal energy can be achieved and this energy is stored by the PCM [84]. By using hexadecanol as PCM encapsulated by PU matrix, very high enthalpy to the order of 229.5 J g^{-1} is possible.

3.6.5. Carbon Based Composites

Carbon based materials such as activated carbon (AC), ordered mesoporous carbon (CMK-5), carbon nanotubes have gained significance, not only as thermally conductive fillers but as containment matrices. Along with improved thermal conductivity of the composite, the porous nature of these materials allows for maximum accommodation of PCM material and offers structural stability without compromising the transition enthalpy significantly. Physical blending and impregnation method have been most commonly applied for preparation of these composites. Due to the porous nature of these materials, there is a minimum threshold for accommodation of working material below which the soft segment gets well absorbed into the pores, thereby hindering the crystallization and thermal behaviour. For the PEG-mesoporous active carbon form-stable PCM, this threshold limit is 30 wt% PEG [85]. The phase change properties of the PEG/AC PCMs are influenced by the adsorption confinement of the PEG segments by the porous structure of AC and also the extent of interference by AC as an impurity with normal crystallization of PEG. The phase change activation energy increases with increasing AC content and decreasing PEG content. For larger chain polyethylene glycol hexadecyl ether (Brij58) working material, this threshold limit is 25 wt% PEG with porous activated carbon supporting framework [86]. Porous AC positively influences the crystallization and nucleation process of Brij58. Confined non-isothermal crystallization of Brij58 as a result of nanoporous activated carbon matrix is associated with increase in half-time and relative degree of crystallinity. The pore structure of these porous materials is also important as it affects the phase change behaviour of composites [87]. The irregular pores of AC have pore diameter of 4 nm whereas hexagonally ordered pores of CMK-5 have pore diameter of 3.5 nm. The mesoporous pores of expanded graphite (EG) have pore diameters around 13 μm. The phase change enthalpies of porous carbon composites and crystallinity of PEG increase in the order of

PEG-AC<PEG-CMK-5<PEG-EG. Thus, higher pore volumes and ordered pore geometries are essential for shape-stabilization. Large pore volume of EG supports favours crystallization, whereas the small pore diameter of AC and CMK-5 hinder the crystallization of PEG.

MWCNT, with (-COOH, -NH$_2$, -OH functional groups) and without functionalization, show different phase change behaviour of PEG in PEG/MWCNT composite [88]. PEG/functionalized MWCNTs exhibit lower phase transition enthalpy and transition temperatures as compared to pure PEG/MWCNT PCM. The functional groups affect the enthalpy values of the composite PCMs in the order of MWCNT-COOH>MWCNT-NH$_2$>MWCNT-OH>MWCNT. This order is a result of the influence of hydrogen bonding, capillary forces and surface adsorption.

Along with direct use of commercially available carbon materials, porous carbon is also synthesized with different raw materials using various routes. By carrying out controlled carbonization of metal organic frameworks at different temperatures, highly porous carbons with exceptionally high surface area and large pore volume reaching 2551 m^2 g^{-1} and 3.1 cm^3 g^{-1} respectively have been used as supporting matrix in PEG-based shape-stabilized PCMs [89]. Carbonization of metal organic frameworks at high temperatures, leading to the migration and evaporation of zinc oxide particles, forms a highly porous matrix with an adsorption capacity of 92.5 wt% PEG which can afford high latent heat of 162 J g^{-1}. Additionally, the thermal conductivity is 50% higher as compared to pure PEG. Potato is also used as raw material to prepare porous carbon by using consecutive freeze drying and heat treatment, thereby utilizing renewable plant resources/agricultural feedstock [90]. The prepared porous carbons exhibit average pore diameter of 204.7 nm with 73.4% porosity, while providing reasonable mechanical strength to retain its shape-stability and prevent leakage of melted PEG. A maximum loading of 50 wt% of PEG can be attained. Melted PEG having good wettability to carbon confirms exudation stability above melting temperatures.

3.6.6. Graphite and Graphite Derivatives Based Composites

The importance of graphite and its derivatives lies in its two-dimensional nature, high thermal conductivity and superior electronic and mechanical properties along with the added advantage of low density. Similar to carbon composites, these composites of graphite are also commonly prepared by using blending and impregnation process. Blends of PEG and EG contain 90 wt% PEG with high enthalpy of 161 J g^{-1} at 61°C [91]. Further, the thermal conductivity improves with increase in percentage incorporation of EG with maximum values reaching 1.324 W m^{-1} K^{-1} at 10 wt% of EG. Three-dimensional netlike architecture of graphene/PEG composites, containing 93 wt% PEG, have fusion enthalpies reaching 166 J g^{-1} [92]. This interconnected architecture of composite is used as a source of heat for thermoelectric device fabrication allowing long steady-state output time with increasing weight percentage of PEG. Another environment friendly derivative of graphene, sulphonated graphene (SG), is used as a nanocomposite matrix to incorporate 96 wt% PEG by using solution processing in aqueous medium [93]. The thermal conductivity increases fourfold from 0.263 W m^{-1} K^{-1} for pristine PEG to 1.042 W m^{-1} K^{-1} for the composite. However, the latent heat value is 12.9% lower than pure PEG.

The oxygen functional groups of graphene oxide (GO) undergo strong hydrogen bonding interactions with PEG up to 90 wt% incorporation [94]. GO is effective in lowering phase change temperature without much impact on phase change enthalpy as compared to other carbon-based porous materials including AC and CMK-5. In a typical procedure, 96 wt% PEG/GO nanocomposites are prepared by using both blending-impregnation and microwave method [95, 96]. The interlayer spacing between GO sheets is enhanced by using microwave, which helps to efficiently disperse the PEG molecular chains. Thus, the TES of the composite prepared by microwave method is 175 J g^{-1}, while the composite prepared by blending and impregnation method has latent heat of 143 J g^{-1}. With increase in concentration of GO, the photo to thermal energy conversion efficiency is improved. The surface of GO can be modified by carboxylation and reduction to prepare GO-COOH and reduced GO respectively; which is used for preparation of PEG based nanocomposites [97]. The carboxylic functionalization leads to a random distribution of carboxyl groups which can interact strongly with PEG chains whereas the reduction with $NaBH_4$ removes surface oxygen functional groups, responsible for weak interactions with PEG chains. Therefore, melting enthalpy and phase change temperature reduction follow the trend of PEG/GO-COOH < PEG/GO < PEG/reduced GO. Apart from surface functionalization, the large interface area is also a contributing factor in the phase change behaviour of PEG. The reduction and functionalization of graphene oxide with oleylamine (OA) (Figure 9(a)) followed by adsorption of palmitic acid and simultaneous self-assembly into three-dimensional structure (Figure 9(b)) results in phase change composite [98]. With very low loading of 0.6 wt% of GO, the composite retains 99.6% of the latent enthalpy of palmitic acid with a value of 196.6 $J\ g^{-1}$. The enthalpy improves with thermal cycling after 500 thermal cycles to 202 $J\ g^{-1}$. The composite exhibits high photo to thermal conversion efficiency.

Three-dimensional hybrid graphene aerogels (HGA) act as both nucleating agent and supporting material. By allowing incorporation of 2 wt% GO and 4 wt% GNP in hybrid graphene aerogels and introducing it in PEG by using vacuum impregnation and physical blending methods, composite PCMs can be made [99, 19]. These composites exhibit high latent heat values in the range of 170-186 $J\ g^{-1}$ and also the three-dimensional supporting framework leads to significant increase in thermal conductivity. The vacuum impregnated composite exhibits an increased thermal conductivity from 0.3 $W\ m^{-1}\ K^{-1}$ of pure PEG to a value of 1.43 $W\ m^{-1}\ K^{-1}$ by incorporating aerogel, having a composition of 0.45 wt% GO and 1.8 wt% GNP. Additionally, this composite PCM shows light to thermal energy conversion with 91.9% efficiency. For the composite prepared by using physical blending method, the thermal conductivity reaches to a maximum value of 1.7 $W\ m^{-1}\ K^{-1}$ and electrical conductivity reaches a value of 2.5 $S\ m^{-1}$. With ultralow filler concentrations, this composite exhibits high latent heat value of 178 $J\ g^{-1}$, which is 98.2% of that of pure PEG.

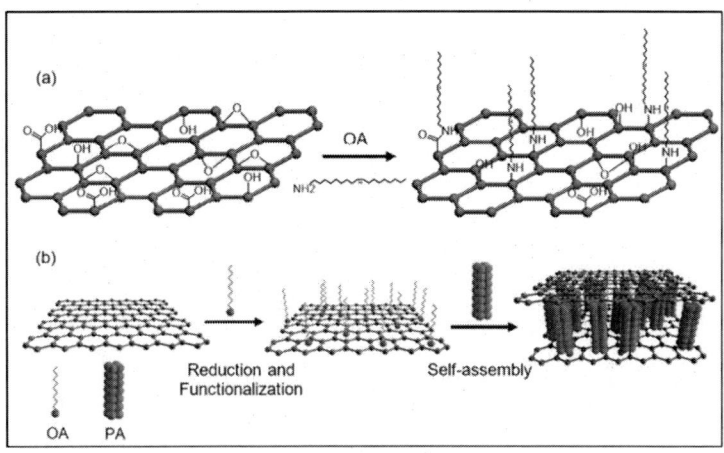

Figure 9. (a) Functionalization of GO and (b) Schematic illustration of assembly of functionalized GO and palmitic acid followed by self-assembly into 3-D composite. "Reprinted (adapted) with permission from J. Phys. Chem. C 119, 40, 22787-22796. Copyright (2015) American Chemical Society."

3.6.7. Phase Change Polymer Blends with Engineering Polymer

Engineering polymers such as epoxy and acrylonitrile offer good chemical resistance, toughness and mechanical strength and therefore, have been used as supporting matrices to prepare form-stable PCMs. Composite of PEG and epoxy resin of diglycidyl ether of biphenyl-A (DGEBA), prepared by casting/moulding contains 75% of PEG-4000 [100]. The epoxy resin with a three-dimensional network acts as the supporting material and imparts mechanical and thermal stability to the PCM. In a similar way, cross-linked polymer matrix of poly(acrylonitrile-co-itaconate) can be blended with PEG up to maximum incorporation of 73 wt% PEG, which has good fusion enthalpy of 119 J g^{-1} [101]. Due to steric hindrance, the movement of PEG is limited leading to reduced size of crystallites and therefore, lower phase transition enthalpies. Due to concern over fire in closed area, fire retardant PCM has come in to consideration. The preparation of halogen free flame retardant form-stable organic PCM can be done by using twin screw extruder technique with paraffin working material and high-density polyethylene as supporting material [102]. With latent heat energy in the range of 79-94 J g^{-1}, these PCMs have improved thermal stability and produce a large amount of charred residue during burning. Highly porous, light-weight, superhydrophobic polypropylene aerogel prepared from industrial wastes are considered as a supporting material to store paraffin [103]. Extremely high PCM loading up to 1060 wt% can be realized with heat storage capacity in the range of 141-160 J g^{-1}.

3.7. Phase Change Polymer Blends with Conducting Polymer

Conducting polymer, polyaniline (PANI) composite with tetradecanol is prepared using in-situ polymerization [104]. This composite has a melting enthalpy of 163 J g^{-1} and possesses the ability to endure heat shock. However, due to processing difficulty, more suitable fatty acids have been preferred for preparing PCM composites with PANI. PANI has been used as a supporting material to encapsulate 82 wt% of myristic acid by using surface polymerization method [105]. Keeping in mind the importance of thermal conductivity, exofoliated GNP has been used as a filler for PANI/palmitic acid composite by using ultrasonication [106]. With a low loading of 7.8 wt% of GNP, high enthalpy of 158 J g^{-1} and thermal conductivity improvement of 237.5% can be achieved.

3.8. Miscellaneous Organic Composites

Many studies have dealt with unconventional hard segments in an attempt to understand the feasibility of various materials as containment matrix. The blends of PEO and potato starch in different ratios of 3:1, 1:1 and 1:3 w/w is held together by the strong intermolecular hydrogen bonding interactions [107]. The solid-solid phase transition is observed for 1:3 and 1:1 w/w PEO/potato starch blend with melting enthalpies 43 J g^{-1} and 47 J g^{-1}, respectively whereas the 3:1 w/w PEO/potato starch blend exhibits solid-solid phase change with partial melting and melting enthalpy of 97 J g^{-1}. For this biodegradable PCM, transition enthalpy is dependent on the strength of hydrogen bonds between PEO and starch. Shape stabilized PCMs of PEG with sugars such as glucose, fructose and lactose allow maximum incorporation of 90 wt% PEG [108]. The blends are held together by hydrogen bonds between ether oxygen and hydroxyl end groups as a result of which the latent heat reduces by only 10-20% as compared to pristine PEG. Both these blends are prepared using cost-effective method and the product is biodegradable in nature.

4. HYBRID FORM-STABLE PCMS

Various inorganic hard segments such as silica, diatomite, montmorillonite, calcium silicate, gypsum, clay, vermiculite and bentonite etc. perform well as supporting materials along with organic soft segments. The incorporation of inorganic carrier material adds to the chemical stability, fire resistance, thermal conductivity and mechanical strength of the composites.

4.1. Silica Composites

Silicon dioxide is a highly promising carrier material due to its high surface area, excellent thermal stability, porous and non-toxic nature. The blend preparation method has been extensively studied using an array of techniques starting from simple solution blending [109], conventional sol gel method [110] to ultrasound-assisted sol gel method [111], temperature-assisted sol gel method [112] and coagulant-assisted sol gel method [113]. All the methods can accommodate around 80 wt% PEG as a result of hydrogen bonding interactions, capillary forces and surface tension forces between PEG and SiO_2.

High melting enthalpy of 138 J g^{-1} is possible with the solution blended PCM, whereas much lower value of 75 J g^{-1} is exhibited by the PCM prepared by conventional sol gel method. Further, the type of sol gel method affects the enthalpy of composite. As an example, a PEG/SiO_2 composite prepared by temperature assisted sol gel method has higher enthalpy of 122 J g^{-1} as compared to that for calcium chloride-assisted composite (91 J g^{-1}) [113]. Complex formation between PEG and calcium ion reduces its crystallization ability and leads to lower enthalpy. For the ultrasonic-assisted sol gel method, using ultrasonic power of 300 W and temperature below 60°C, good phase change enthalpy of 103 J g^{-1} can be realized [111]. Low molecular weight PEG/SiO_2 composite exhibit poor thermal behaviour due to confinement of PEG chains by silica framework [110]. The enthalpy and crystallization behaviour of composite improves with increase in molecular weight of PEG used in the composite preparation.

Waste raw materials are also used for making form stable PEG composite. Oil-shale ash is used to extract sodium silicate by using calcination-alkali leaching method under optimized conditions [114]. The PEG-SiO_2 composite, prepared by using temperature-assisted sol-gel method in the absence of surfactant or co-solvent, exhibits maximum latent enthalpy of 152 J g^{-1} at 58°C. The supercooling extent, melting and solidifying time are lower by 22, 27 and 23% respectively than that of pure PEG.

The enhancement of thermal conductivity is a prerequisite for practical application of PCMs. The thermal conductivity of PEG/SiO_2 composites can be enhanced by certain metal and carbon-based fillers. The addition of β-aluminium nitride enhances the thermal conductivity from 0.3847 to 0.7661 W m^{-1} K^{-1}, while the loading is varied from 5% to 30 wt% [115]. Together with increase in thermal conductivity, the enthalpy values decreases from 161 to 130 J g^{-1}. Copper can also be introduced in-situ via chemical reduction of copper sulphate by applying ultrasound-assisted sol-gel method [116]. With 2.1 wt% copper, the thermal conductivity of the composite reaches 0.414 W m^{-1} K^{-1} with corresponding melting enthalpy of 110 J g^{-1}. The introduction of graphite to the PEG-SiO_2 composite enhances the thermal conductivity substantially (0.558 W m^{-1} K^{-1} with 2.7 wt% graphite) [112]. By dispersion of MWCNT in PEG/SiO_2 composite, thermal conductivity can be improved (0.463 W m^{-1} K^{-1} with 3 wt% MWCNT) [117]. The presence of MWCNT qualifies the PCM to exhibit high photo-to-thermal conversion.

4.2. Mesoporous Matrices

Mesoporous matrices are attractive supporting materials due to their unique pore structures, high adsorption capacities, surface permeability and fire-retardancy. The phase change properties and shape stabilization factor are directly related to the average pore size of the material [118]. If the pore size is too small, then PCM crystallization will be impeded and if the pores are large, then the capillary force might not be sufficient to hold the liquid PCM.

Simple physical blending and impregnation process have been utilized in the preparation of composites of both PEG and its derivatives (polyethylene glycol octadecyl ether and polyethylene glycol hexadecyl ether) with mesoporous silica [119, 120]. PEG/silica PCMs are bound by a combined effect of hydrogen bonding, capillary forces and surface adsorption phenomenon to exhibit higher enthalpy w.r.t. mesoporous silica molecular sieves (MCM-41 and SBA-15) [119].

Radial mesoporous silica (RMS), prepared with cetyltrimethylammonium bromide template and tetraethylorthosilicate (TEOS) as SiO_2 precursor, can be vacuum impregnated with PEG [121]. Immersion time of 50 minutes with an immersion temperature of 70°C is optimum for the preparation of this PCM. Maximum enthalpy of 130 J g^{-1} can be realized by using 80 wt% PEG/RMS composite. The extent of supercooling reduces significantly by 4.8% - 19.7%.

Mesoporous silica nanoparticles including MCM-41, SBA-15 and mesocellular foams (MCF) composites with lauric acid is prepared using evaporative solution impregnation to achieve dual thermal responsive PCMs [122]. MCF, with high pore volume can accommodate 83 wt% lauric acid and had enthalpies reaching up to 124 J g^{-1}. The dual temperature response of MCF is possible due to nanoconfined phase of organic material inside mesoporous material and interparticle voids filling on external silica surface. Hexagonal ordered silica viz. MCM-41 and SBA-15 have high fraction of mesopore volume occupied by non-melting layer and are therefore, less likely to have dual temperature response

(Figure 10). This dual temperature range is helpful in achieving large operating window and to improve heat transfer rate.

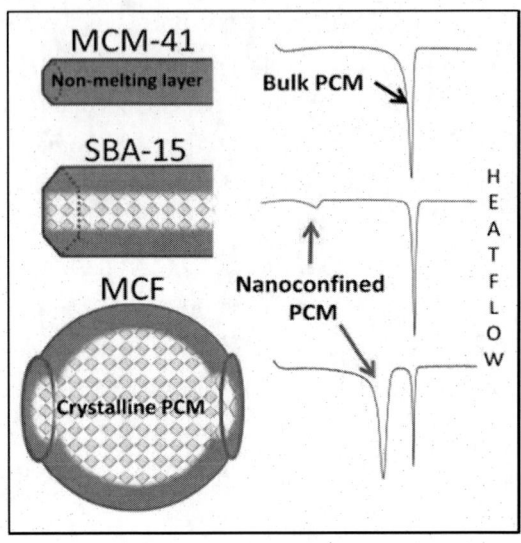

Figure 10. Dual temperature response by mesocellular foams as compared to hexagonally ordered mesoporous silica. "Reprinted (adapted) with permission from J. Phys. Chem. C 119, 27, 15177-15184. Copyright (2015) American Chemical Society."

The modification of surface functional groups of mesoporous silica supporting material, namely SBA-15 can be engineered to control the thermal properties of composite and crystallization behaviour of PCM [123]. The terminal functional groups of PEG/HO-SBA-15-OH are modified with amino and methyl groups to yield PEG H_2N-SBA15-NH_2 and H_2N-SBA15-CH_3, respectively. The amino groups are fabricated to the internal surface of the SBA-15 channels whereas the methyl groups are placed on the external surface, thereby inducing opposing polarities, which prevent the spilling of PEG from SBA-15 channels and promote crystallization of PEG. The grafting of $-NH_2$ groups in channels of silica reduces the adsorption sites of PEG chain and changes the adsorption confirmation from train to loop structure, effectively enhancing the crystallization of PEG. The fusion and crystallization enthalpies of PEG/H_2N-SBA15-CH_3 reach 88 J g^{-1} and 82 J g^{-1} respectively with 70 wt%

PEG. The PEG/HO-SBA-15-OH composite, having only –OH grafting, cannot crystallize effectively and therefore, has zero enthalpy. Similarly, with the modification of surface functional groups of MCM-41 silica support, good enthalpy of 80 J g^{-1} is possible with PEG. Thus, the surface functionalization behaviour of various silica supports can be modified effectively to enhance the crystallization of PEG.

4.3. Building Material Composites

The fabrication of composite PCMs by using porous building materials as a supporting matrix for the incorporation of liquid PCM, offers the advantages of thermal comfort and energy efficient maintenance of indoor temperature. Simple blending-impregnation and vacuum impregnation processes have been utilized for the preparation of building composite PCMs with maximum incorporation of soft segment.

Both ordinary portland cement [124] and mesoporous calcium silicate (MCS) [125] can be impregnated with PEG using simple physical blending process, allowing an incorporation up to 70 wt% of soft segment. The PEG molecules are confined in the pores as a result of the combined effect of surface tension and capillary forces. Vacuum impregnation of gypsum and natural clay can accommodate a maximum of 18 and 22 wt% PEG, respectively and therefore, yielding low enthalpies of 24 J g^{-1} for PEG/gypsum and 29 J g^{-1} for PEG/clay composite [126]. While studying thermal performance of these building PCMs, by fabricating wallboards with both the composites using cubicle systems, the inside centre temperature of cubicles made of PEG/natural clay and PEG/gypsum are found to be lower by 2.08°C for 60 minutes and 1.47°C for 120 minutes, respectively as compared to outside temperature. Low-cost clay mineral, bentonite exhibits melting temperature and melting enthalpy varied in the range of 4-30°C and 38-74 J g^{-1} respectively when its blend with capric acid, PEG 600, dodecanol and heptadecane are prepared by vacuum impregnation [127]. The transition temperature range is suitable for passive solar heating and cooling in building envelopes. By adding 5 wt% EG, the thermal conductivity of bentonite-capric acid, bentonite-PEG 600, bentonite-dodecanol and bentonite-heptadecane composites can be increased by 65%, 63%, 39% and 47% respectively. Also, the heating times of the bentonite-PCM composites with EG, can be reduced by 12-22 seconds.

Natural Na-MMT clay, having average size of 936-1190 nm, is used to produce PEG-MMT composites using ultrasonic impregnation, incorporating up to 44 wt% of PEG [128]. With the original heat storage capacity of 104.8 J g^{-1} remaining almost constant till 100 thermal cycles, these composites are considered as promising candidates for the design of thermal insulation systems. The montmorillonite nanoclay can be used only for improvement of thermal properties of the nanocomposite over neat PEG/epoxy blend [129]. The presence of nanoclay maintains the heat transfer barrier properties, which is manifested by a 31% reduction in top surface temperature.

4.4. Composites with Diatomite

Another economical, porous building material, diatomite has gained significant attention as a hard segment for preparation of solid-solid PCM due to its chemical inertness, high surface area, high adsorption capability and low density. It can be directly incorporated [130] or vacuum impregnated [131-132] to prepare composite PCM. For preparation of composite by direct impregnation method, an immersion time of 90 minutes at immersion temperature of 75°C is ideal [130]. A loading of 55 wt% PEG can be realized with transition enthalpies ranging from 78–107 J g^{-1}. By using vacuum impregnation method, maximum incorporation of 58 wt% PEG can be achieved [132]. As an example, the phase change enthalpy of PEG 4000/diatomite PCM is 106 J g^{-1} with a transition temperature of 58°C, which is highly stable after repeated thermal cycles. PEG/diatomite/single walled carbon nanotubes (SWCNTs) form-stable composite PCMs, containing PEG as high as 60 wt%, exhibit melting point shift to a lower temperature while the solidification point shifts to a higher temperature due to the presence of SWCNTs [133]. The melting and solidification time of SWCNTs loaded matrix decreases by 54.7% and 51.1%, and thermal conductivity increases by 2.8 times as compared to the composite PCM without SWCNTs. No leakage occurs by extensive thermal cycling study.

Raw diatomite is normally leached with alkali to dredge the pores of diatomite and thereby, improve the surface area and pore size [134]. Spherical, crystalline silver nanoparticles (Ag-NP) (3-10 nm) can be uniformly deposited on the surface of diatomite, which is blended with PEG to enhance the heat transfer properties of PCM. About 63 wt% PEG can be loaded to exhibit melting enthalpy of 111 J g^{-1} at 60°C. With 7.2 wt% of Ag, a thermal conductivity of 0.82 W m^{-1} K^{-1} can be achieved, which is 127% higher than PEG/diatomite composite.

4.5. Miscellaneous Hybrid Composites

Among the hybrid composites, a number of unconventional inorganic hard segments have been utilized to prepare solid-solid PCMs with a wide array of properties.

Phosphamide containing silsesquioxane matrix, with an average pore diameter of 4.1 nm, is utilized as supporting matrix for preparation of fire-retardant PCM. With 82 wt% PEG, the melting enthalpy can be as high as 125 J g^{-1} at a melting temperature of 56°C [135]. Bischofite, a by-product of non-metallic mining industry, is obtained during potassium chloride production with a major composition of $MgCl_2 \cdot 6H_2O$ [136]. $MgCl_2 \cdot 6H_2O$ itself is a well-recognized inorganic PCM. Its cycling performance can be most satisfactorily improved by adding 5% PEG. However, high supercooling of 37°C remains a disadvantage.

Bulk carbon nitride (bulk-C_3N_4) and carbon nitride intercalation compounds (CNIC) can be used as shape-stabilizers for PEG by using blending and impregnation method [137]. The higher specific area of CNIC in PEG/CNIC blends than that of bulk C_3N_4 aids the crystallization of PEG. The PEG/CNIC blend possesses a fusion enthalpy of 46 J g^{-1} for

60 wt% PEG with a reduction of melting temperature by 24°C from that of pure PEG. Thus, the use of graphitic carbon nitride as shape-stabilizers helps in lowering the phase change enthalpy and extent of supercooling.

To PEG/expanded vermiculite composite, silver nanowire is added as thermal conductivity enhancer [138]. The silver nanowire having length 5-20 µm and diameter 50-100 nm, is wrapped by PEG and is well-infused into the pores of expanded vermiculite. A maximum of 66 wt% PEG can be utilized for shape-stabilization. On the other hand, increasing concentration of silver nanowire decreases the latent enthalpy reduces and increases the thermal conductivity (enthalpy of 99 J g^{-1} and thermal conductivity at 0.68 W m^{-1} K^{-1} at 19.3 wt% Ag nanowire). Expanded vermiculite acts as a nucleating agent in promoting the crystallization of PEG chains and causes reduction of supercooling extent by 7°C. By adding silicon carbide nanowires as conductive fillers to vermiculite composite, the conductivity can be enhanced to 0.53 W m^{-1} K^{-1} with a melting enthalpy of 65 J g^{-1} at 3.28 wt% of the former [139]. Hybrid composite PCMs can be prepared by using inherently thermally conductive inorganic supporting materials, such as flower-like TiO$_2$ nanostructures with unique pore structure and high surface area that is obtained by hydrothermal method, which contains 50 wt% PEG with melting enthalpy of 86 J g^{-1} [140]. With 4 wt% GO and 30 wt% boron nitride serving as hard segments and thermally conductive fillers, PEG can be constricted to undergo crystalline to amorphous phase transition without liquid leakage [141]. The melting enthalpy and thermal conductivity of the composite are 107 J g^{-1} and 3 W m^{-1} K^{-1} respectively and have the capabilities to exhibit light to electric energy conversion.

5. CHEMICAL METHODS FOR ACHIEVING SOLID-SOLID PCM

Various chemical methods such as grafting, cross-linking and copolymerization can be utilized for the preparation of solid-solid PCMs, wherein the working material is chemically linked to the supporting material via different linkages such as urethane, epoxy, ester and ether modification. The chemical methods of modification yield more chemically and thermally stable PCMs but compromise the TES capacity.

5.1. Polyurethane Modification

Among the chemical methods, urethane modification is most commonly applied for preparation of chemical solid-solid PCMs. A wide array of multifunctional chain extenders with methylene diphenyl diisocyanate (MDI) have been utilized for the preparation of polyurethane PCM including 1,4-butane diol [142], pentaerythritol [143], polyvinylalcohol [144], β-cyclodextrin [145], trihydroxy surfactants [146] and hexahydroxy polyols [147]. Undergoing a crystalline to amorphous phase transition with reduction of crystal size of PUPCM compared to that of pure PEG, there is also an increase in thermal resistance due to cross-linking and incorporation of phenyl groups. In SSPCMs with low crosslink density, the PEG soft segments are less hindered to move freely and maintain higher crystallization ability [145].

Thermoplastic polyurethane exhibiting solid-solid phase transition can be prepared by using PEG as soft segment and MDI along with novel tetrahydroxy compound as hard segment [148]. The lower phase change enthalpy of 137 J g^{-1} at 56°C is due to lower crystallization ability of PUPCM than pure PEG. However, these materials are thermally stable with good mechanical properties and processing abilities. PU SSPCM is synthesized by two step condensation process by reaction of PEG and MDI with novel tetrahydroxy compound. Terephthalic acid bis-(2-hydroxy-1-hydroxymethyl-ethyl) ester can also be used as tetrahydroxy compound, which can be reacted with PEG and MDI to get PUPCM having phase change temperature of 48°C and latent heat of 154 J g^{-1} [149].

By employing PEG as soft segment with isophorone diisocyanate (IPDI) and BDO as hard segments, PEG-PU shape memory polymer can be prepared, which has enthalpy value of 100 J g^{-1} [11]. The hydrogen bonded hard segments restrict the movement of the soft segments when heated above the phase transition temperature of 47°C. The PU prepared with PEG 1000 and coupling reagents, IPDI and TDI exhibits solid-liquid PCM behaviour and all other polyurethane PCMs (PEG mol. wt. 6000, 10000) possess solid-solid transition behaviour [150]. Due to linear and symmetric backbone, PEG-HMDI PCMs have higher phase change enthalpies than PEG-IPDI and PEG-TDI PCMs. With higher molecular weight PEG (PEG 6000 and PEG 10000), the transition temperatures remain unaffected with variation in diisocyanates type. The enthalpy values are significantly lower and overcooling is more pronounced for low molecular weight PEG (PEG 1000). The melting enthalpy and thermal efficiency of the biodegradable polymeric PCM from HMDI/PEG/castor oil combination is better than the PCM with MDI due to the steric hindrance caused by the rigid benzene rings of MDI [151]. Novel comb-polyurethane can be prepared by reaction of soft segment, monomethoxy PEG modified by diethanolamine with hard segment consisting of IPDI and BDO [152]. As the soft segment is linked to the main PU chain only by one end compared to the restriction of chain from both ends in PEG-PU, the comb-PU shows higher phase transition enthalpy of 122 J g^{-1} and better crystallization properties than PEG-PU.

Two different PEG based components viz. isocyanate terminated prepolymer and tetrahydroxy terminated prepolymer can be crosslinked to ease the processing of thermosetting SSPCMs in solvent-free medium, resulting in the formation of flexible PCM [153]. Phase change enthalpy of 98 J g^{-1} is exhibited by the flexible PCM having thermal stability higher than 300°C. Ionic polyurethanes are synthesized by using PEG as soft segment with MDI, N-methyldiethanolamine (MDEA) and 1,3-propanesulfonate as the hard segments [154]. This PCM exhibits high phase change enthalpy of 152 J g^{-1} when PEG 10,000 (ΔH_m= 187 J g^{-1}) is used for forming the ionomer. The introduction of ionic groups results in better phase separation, improved thermal stability and reduced interference of hard segments during crystallization of soft PEG segments. Using a grafting-to method, PVA-grafted-octadecanol copolymers are prepared using urethane bonds. Compared to pure octadecanol, the melting enthalpy is much reduced for the copolymer but can be improved with increasing grafting ratio [155].

Hexamethylene diisocyanate biuret (HDIB), crosslinking agent and hard segment are reacted with PEG along with addition of halloysite nanotubes [156]. These nanotubes act as crosslinking and nucleating agent increasing the rate of crystallization and hence, thermal energy storage density of the system. The excessive crosslinking by HDIB leads to reduction in phase change enthalpy, which is improved by the presence of halloysite nanotubes due to the heterogeneous nucleation effect enhancing to a value of 118.7 J g^{-1}.

5.2. Hyperbranched Polymers

Hyperbranched polyurethanes are being explored as solid-solid PCMs due to their many advantages including easy synthesis, large number of reactive end groups, highly branched structure, minimal chain entanglement and maximum chain rearrangement. Initial reports point towards the preparation of hyperbranched polyurethane as solid-solid PCM using PEG, diisocyanate and hyperbranched polyester as a chain extender [157-160]. Due to the presence of strong covalent bonds, the hard polyurethane segment prevents the melting of PEG to liquid state during crystalline to amorphous phase transition [158]. The hyperbranched polyurethanes maintain a solid-solid phase transition when the molecular weight of PEG is maintained beyond 2000 [159]. Also, the phase transition temperatures and enthalpies reduce with reducing soft segment concentration and increase with increasing soft segment molecular weight. The strategy of doping the hyperbranched polyurethane with pentaerythritol improves the phase change enthalpy and thermal resistance of the material [160]. PEG is also used in the preparation hyperbranched polyol which is used later as a chain extender [161]. For easy synthesis and processing in large scale, hyperbranched polyurethane can be synthesized from PEG soft segment and trifunctional aromatic (phloroglucinol) (Figure

11) and aliphatic (trimethylolpropane) core moieties as branching units [34, 162].

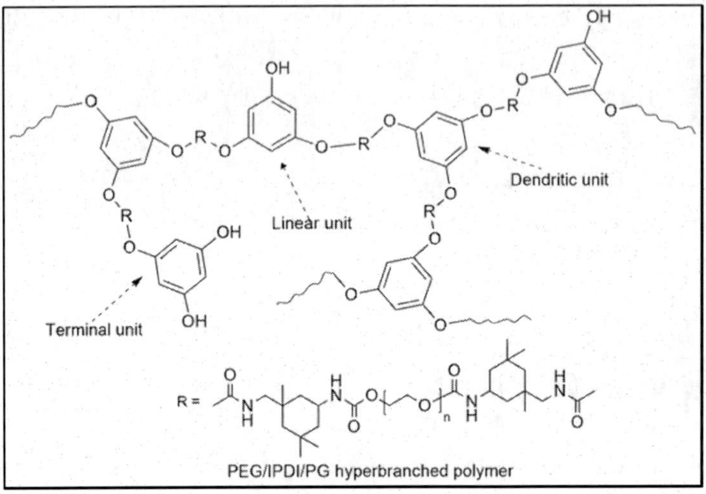

Figure 11. Structure of hyperbranched polyurethane with aromatic branching core unit, phloroglucinol. "Reprinted with permission from Ind. Eng. Chem. Res. 56, 49, 14401-14409. Copyright (2017) American Chemical Society."

5.3. Natural Polymer Grafted SSPCMs

Natural polymers impart properties such as biodegradability, renewability, biocompatibility, and non-toxicity, which make them attractive for use as hard segments in the preparation of solid-solid PCMs. Cellulose from MMTritylcellulose can be used to prepare cellulose-g-PEG copolymers using ether linkage with thermal storage capacity of more than 120 J g^{-1} and phase transition varying from room temperature to 50°C [163]. The PCM undergoes solid-solid transition only when cellulose composition is more than 15%. Graft copolymers with polyethylene glycol monomethyl ether (mPEG) soft segment and cellulose hard segment are prepared via formation of urethane bonds in ionic liquid [164]. The thermal energy storage capacity can reach 154 J g^{-1}. Cotton with major cellulose constituent is also bound to PEG with urethane linkage [165]. With a grafting percentage of nearly 30%, the material has potential for the development of smart fabric as the inherent properties of cotton cloth remain mostly unchanged after grafting and the resistance to degradation increases during wash. Derivative of cellulose, cellulose diacetate (CDA) based PCM with PEG is prepared by urethane bonding [67]. PEG molecular weight of 10,000 exhibits maximum enthalpy.

5.4. Modified PEG Polymers

PEG derivatives are appealing as solid-solid PCMs due to their similar crystal structure and crystallization behaviour as PEG along with the added advantage of designing polymerizable monomers. These derivatives of PEG also have high phase change enthalpy and adjustable transition temperature range over a wide range of molecular weights. Copolymerization reactions have been used to prepare PEG based copolymers with parallel thermal energy storage capacity. Styrene can be copolymerized with vinylic end group of PEG monomethyl ether yielding TES material with maximum storage capacity of 109 J g^{-1} at phase change temperature of 58°C [166]. The crystalline properties of PEG monomethyl ether remain unaffected during synthesis. Polydecaglycerol acrylate and PEG acrylate are copolymerized together to prepare SSPCM having melting enthalpy of 163 J g^{-1} [167]. High enthalpy value is facilitated with extensive hydrogen bonding from polydecaglycol chains. Poly(polyethylene glycol methyl ether methacrylate) homopolymer can be prepared via free radical bulk polymerization of polyethylene glycol methyl ether methacrylates, which exhibit good crystallinity [168].

PEG acrylate based crosslinked copolymer can be synthesized to enhance the thermal stability, while retaining high value of enthalpy. This is possible as PEG is present as side chain. The crosslinked copolymer PCMs possesses thermal stability beyond 300°C with maximum fusion enthalpy of 145 J g^{-1} [169]. PEG acrylate can be copolymerized with methyl methacrylate to realize maximum enthalpy of 165 J g^{-1}. The monomer reactivity ratio calculation indicates strong tendency to undergo cross-propagation [170]. PEG octadecyl ether monomer can be synthesized with varying the PEG chain length, which has transition enthalpy varying in the range of 143-183 J g^{-1} [171]. Polymeric PCM can be prepared by copolymerization of PEG octadecyl ether monomer with methyl methacrylate to prevent the leakage of the liquid phase [172]. Copolymer poly (PEG alkyl ether vinyl ether) is prepared with varying lengths of PEG spacers having highest enthalpy of 103 J g^{-1}. The PCM is thermally stable up to 295 °C and thus, can be used for preparation of melt-spinning thermoregulated fibre.

5.5. Miscellaneous Polymeric Supports

Using diverse polymeric supporting materials SSPCMs can be synthesized chemically. Polyethylene terephthalate (PET)-PEG block copolymers [173] and coupling blend of PEG with polyacrylamide [174] can be synthesized using chemical methods. PEG with carboxyl groups can be linked with Eu^{3+} to prepare phase change luminescent material with phenanthroline as secondary ligand [175]. The Eu-PEG polymeric SSPCM has phase change enthalpy of 97 J g^{-1}, having luminescent properties with UV absorption at 309 nm.

PEG modified with maleic anhydride can be grafted onto polyacrylonitrile chain yielding comb graft architecture SSPCM with phase change enthalpy of 74 J g^{-1} [176]. PEG can be grafted to poly(styrene-co-acrylonitrile) matrix for preparation of solid-solid PCM [177]. The architecture has thermal stability up to 350°C and thermal reliability beyond 1000 cycles. Form-stable phase change materials of cross-linked poly(acrylonitrile-co-itaconate)/polyethylene glycol (PEG) blend exhibits thermal reliability and heat storage durability after 1000 thermal cycles [100]. The phase change temperature, being in the range of 23–53°C with good phase change enthalpy of 119 J g^{-1}, the material is well suited for commercial applications.

Cross-linking reaction of PEG with melamine-formaldehyde [178] and carboxylated PEG with epoxy groups of poly(glycidyl methacrylate) [179] results in PCM, wherein the hard segment restricts the flow of working material resulting in low enthalpy. Polystyrene-graft-PEG copolymers with different mass percentage of PEG content exhibit latent heat values ranging between 116-174 J g^{-1} with transition temperature varying from 55 to 58°C [180]. Palmitic acid is grafted onto polystyrene backbone using palmitoyl chloride to prepare solid-solid PCMs with a maximum incorporation of 75 wt% palmitic acid [181]. Although, the graft polymer has a low storage enthalpy of 40 J g^{-1}, the thermal reliability for 5000 thermal cycles is quite high. Environmentally friendly PCM containing poly(3-hydroxybutyrate-co-3-hydroxyvalerate) *(PHBV)*/PEG copolymer had latent heat of fusion in the range of 126-135 J g^{-1} above 20% PHBV fraction [182]. Three-dimensional semi-interpenetrating polymer networks can be prepared by using simultaneous interpenetrating technique [183]. One network is formed by condensation polymerization of TEOS and PEG whereas the second network is prepared by free-radical polymerization of 2-hydroxyethyl methacrylate (HEMA) to form poly(HEMA). Although interpenetrated, high transition enthalpy of 145 J g^{-1} can be achieved. Further, the relative hydrophilic nature of PEG can be reduced with increasing HEMA concentration.

6. APPLICATIONS OF PCMS

Phase change polymers are being evaluated for large number of applications and many have been inducted in to commercial use as maintenance free heating/cooling system. The possible application areas are air conditioning, telecom shelters, as the interruption of power supply can be taken care by PCM. Transportation of perishable foods, temperature sensitive pharmaceuticals, sundry electronics and chemicals (explosives) require cooling, which can be done by PCM at much cheaper rate as compared to conventional methods. Automobiles sector has tremendous application of PCM for better utilization and comfort. In construction, PCMs are embedded to make it more thermally comfortable than it is without PCM. Heating or cooling houses in different climate can be done with PCM to great extent. PCM can be used for keeping caterer's food hot during transportation. Greenhouse is another area, where PCM can be comfortably used for maintaining temperature variation at minimum level.

6.1. Textile Industry

Almost 30 years ago, PCM was first brought into application by National Aeronautics and Space Administration (NASA) for use in spacesuits to protect the astronauts from temperature fluctuations in outer space. Thus, the concept of smart textiles for human comfort came into existence in the form of parkas, gloves, sports apparel, footwear and thermal undergarments. These smart textiles can regulate the body temperature by controlling the heat uptake or release depending on variation in external environment. The incorporation of PCM into fabric can be performed by coating, impregnation of hollow fibres with PCMs, microencapsulation or electrospinning.

During the early years, various fabrics including polyester, wool, nylon 66 and cotton were treated with aqueous solutions of PEG using conventional pad-dry technique to enhance the thermal properties by minimum of 2-2.5 times than that of untreated fabric [184]. The interactions between PEG and the amide groups of nylon and wool are the reason for exhibiting highest enthalpy among various fabrics. The thermal values are also quite stable even after 50 thermal cycles.

PEG nanocapsules with urea-formaldehyde shell are fixed onto cotton fabric by using pad-dry-cure method [185]. Increasing the binder agent improved the tensile strength and abrasion resistance of the treated fabric.

Solid-solid PCM can be produced by grafting PEG 1000 onto cotton waste fibres with glutaraldehyde as cross-linker [186]. A difference of 1-1.5°C in temperature is observed for the box containing the PCM for 23-25 min. Poor adhesion properties of PEG to non-cellulosic fibre [poly(lactic acid)] is improved by using highly cross-linked PEG system consisting of polyethylene glycol-dimethyloldihydroxyethyleneurea (PEG-DMDHEU) [187]. This strategy helped in improving wash fastness while the enthalpy value is mostly unaltered. The PLA fabric shows a weight gain of 44% after cross-linking, allowing less air to permeate through the fabric, improve the thermal properties, better static charge dissipation and lower surface tension as compared to the PET fabric used. The improvement in adhesion of PEG to wool fibres is done by using fluorinated and Chlorine-Hercosett pre-treated wool fabrics followed by reaction with PEG-DMDHEU [188]. The fluorinated fabric shows highest thermal properties with melting enthalpy value of 17 J g^{-1} for washed fabric together with better dimensional stability, enhanced felting performance, reduced water vapour permeability and increased air permeability.

Non-woven mats comprising PEG core and PVDF sheath are prepared by coaxial electrospinning [189]. The use of non-woven mats adds to the mechanical strength and specific surface area with desirable dimensions over conventional electrospun fibres and eliminates the need for any further fibre processing. The optimized nanofibers, having average diameter of 721 nm and latent fusion enthalpy of 34 J g^{-1}, has encapsulation ratio of 20 wt% PEG. These fibres exhibit a reduced degree of elongation with introduction of PEG into PVDF, but satisfactory mechanical strength is retained. To improve the mechanical strength of the PEG-PVDF nanofibers, fumed silica (2 wt%; PEG 1000) is added to the spinning solution [190]. The diameter of the nanofibers reduces with an increase in fusion enthalpy to 59 J g^{-1} and improvement of tensile strength.

Flexible, porous membrane can be prepared by dispersing two PEGs of different molecular weights into polyurethane solution [191]. The membrane preserves its solid state on heating above the PEG transition temperature with enthalpies reaching up to 129 J g^{-1}. The PU/PEG flexible membrane has porous structural, suitable transition temperature and high transition enthalpy. The porous technology introduction into functional textile's formation contributes to the development of a new way to improve the phase change enthalpy largely for adjustable textile.

Direct grafting of PEG onto cotton framework, as indicated by POM images in Figure 12, is done up to a grafting percentage of 30 wt% [165]. The grafted cotton has improved tensile strength along with an increase in gram per square metre value. A reduction in tear strength and increase in abrasion resistance from 33,000 rubs to 42,000 rubs for thread breaking has direct relation with strong chemical bonds of PEG onto cotton. The water, perspiration and rubbing fastness of the fabric remains almost similar.

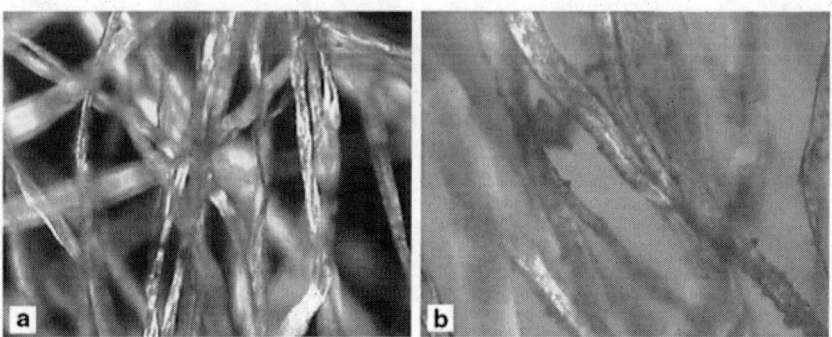

Figure 12. POM images of (a) pure cotton and (b) modified cotton. The roughness in morphology in Figure 6(b) is indicating grafting of PEG and therefore, the modified fibre can directly be applied for textile production.

6.2. Building Applications

With increasing demands for comfort levels inside the living environment, buildings have been found to contribute at least 40% of total world energy consumption and therefore, among one of highest energy consuming sectors. PCM wallboards, shutter, underfloor heating system, ceiling board, heat exchanger etc. find application for heating or cooling of the building [4]. The PCM can be applied by adding PCM filled pellets to wallboards during manufacturing or by impregnating wallboards with PCM, using PCM enhanced concrete or PCM incorporated PU foam [192].

Diatomite or diatomaceous earth is a natural amorphous silicate possessing high porosity, excellent absorption capacity, chemical stability and low cost [131]. This lightweight building material can be integrated with a maximum of 50 wt% PEG to prepare shape-stabilized PCMs. The addition of EG improves the heat transfer rate of the PEG/diatomite composites. The wallboards fabricated with composite PCM, exhibit a temperature difference of 2.3°C on the inner surface.

PEG-cement composites are utilized directly for building applications [124]. When the mass ratio of PEG and cement is raised to 1:3, the corresponding fusion temperature and enthalpy of the PCM are 24°C and 24 J g^{-1}. PEG is well-impregnated into porous cement showing no leakage during phase transformation. In the PEG-MCS shape-stabilized composite, 70 wt% of PEG can be accommodated resulting in melting and solidification enthalpies of 122 and 107 J g^{-1} and the corresponding heat storage and heat release are improved by 28.2% and 27.3% respectively as compared to pure PEG [125]. The direct impregnation of PEG into building material, MCS with transition temperature in the range of 50-70°C makes it suitable for building envelopes during hot summers.

Porous building materials, gypsum and natural clay are used as supporting materials for preparation of PEG containing shape-stabilized composites using vacuum impregnation method [126]. A maximum loading of 18 and 22 wt% PEG in PEG/gypsum and PEG/natural clay composite can be realized, respectively. The thermal performance of wallboards fabricated from PEG/natural clay and PEG/gypsum PCMs, are able to maintain temperature difference of 2.08°C for 60 min and 1.47°C for 120 min, respectively. Low-cost clay mineral, bentonite, impregnated with four different kinds of PCMs including capric acid, PEG 600, dodecanol and heptadecane, exhibits wide range of melting temperature and melting enthalpy (4-30°C and 38-74 J g^{-1} respectively) [127]. The transition temperature chosen makes it suitable for passive solar heating and cooling in building envelopes. Thermal conductivity is also increased by adding EG (5 wt%), so that response times are lowered.

6.3. Solar Energy Storage

Solar energy is the most abundantly available form of renewable energy. Several techniques have been explored and utilized for effectively capturing solar radiation. Solar energy conversion to thermal energy is one of the most efficient methods to capture large quantity of thermal energy in a small volume. The low thermal conductivity of organic PCMs is a major drawback in its application, but the preparation of efficient TES materials with high thermal conductivity and high solar energy absorption efficiency are being developed.

A PEG based solid-solid PCM is developed with organic dye, 1,4-bis((2-hydroxyethyl)amino)anthracene-9,10-dione as supporting framework for functioning as solar energy harvester [31]. The PEG 20000-dye PCM exhibits melting enthalpy of 143 J g^{-1} and absorption in the visible range at 555 nm and high light to thermal conversion with an efficiency of 0.937. Under sunlight irradiation, the temperature of the composite increases to 70°C in 34 min thereby, effectively absorbing visible sunlight through non-radiative process (Figure 13).

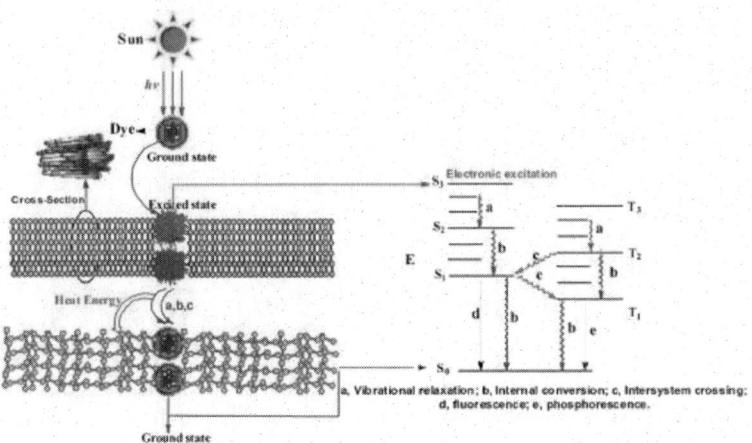

Figure 13. Proposed mechanism for solar-thermal conversion by PEG incorporated organic dye through nonradiative decay and generated latent heat is stored by PCM. Reproduced from ref. Wang, Tang, and Zhang 2012b with permission from the Royal Society of Chemistry.

The heat supply for thermoelectric device is generated by using graphene/PEG composites with a three-dimensional netlike architecture [92]. This interconnected architecture enables efficient heat transfer through thermally conductive pathways. Graphene/PEG composites have superior thermal and electrical conductivity together with high structural stability. High PEG loading of 93 wt% and corresponding high melting enthalpy of 166 J g^{-1} with 93.3% efficiency ensure longer steady-state output time and therefore, higher heat output is generated by the thermoelectric device.

For effective light harvesting, MWCNTs can be blended with PEG-SiO$_2$ to make composite having improved thermal conductivity, absorption over a broad range of sunlight, high light to heat conversion and energy storage efficiency [193]. The maximum latent heat with 0.5 wt% of MWCNT remains 139 J g^{-1} at 54°C and thermal conductivity at 0.389 W m^{-1} K^{-1}. The efficiency of light-to-heat conversion and thermal storage is very high at 0.918.

Single-walled carbon nanotube (SWCNT) (dia: < 2 nm, length: 5-20 μm), modified by introducing nitrophenyl groups and stabilized in toluene, can be added to form-stable PCM derived from PEG, diisocyanate and N, N'-dihydroxyethyl aniline [194]. SWCNT functions as an effective photon capturer and shows light to thermal energy conversion with energy storage efficiency ranging from 0.845-0.913 upon solar irradiation. The enthalpy is higher than 100 J g^{-1} above 50°C and thermal conductivity is higher by 25.1%, which results in low heating time of 870 sec. Due to strong absorbance in near IR zone, these thermally reliable and stable composite PCMs can be used in military stealth and smart textile applications. Smart energy storage composite having efficient electro/photo to heat conversion and storage with high thermal conductivity can be developed by cross-linking copolymerization of PEG and IPDI after injecting into graphite foam [195]. This material is found to have thermal conductivity of 3.5 W m^{-1} K^{-1}. By incorporation of more than 80 wt% PEG 8000 in the graphite foam, the electro to heat storage efficiency can be enhanced to a maximum of 80% at low voltages and the photo to heat conversion efficiency of 67%.

The 3D heat conductive pathway formed in PEG10000/HGA composite (GO and GNP amount varied) results in high thermal conductivity of 1.43 W m^{-1} K^{-1}, high fusion enthalpy ranging from 178-182 J g^{-1} and crystallization enthalpy from 170-174 J g^{-1} [16]. The energy storage efficiency from solar simulator (xenon lamp) reaches 80-92%

using tangential method and effective light-to-heat conversion is achieved within 33 min.

6.4. Innovative Applications of PCMs

The conventional applications of PCMs for thermal energy storage have been recognized widely and commercially applied by many companies such as Outlast technologies, Rubitherm, Dupont, Microtek, Sunamp, Teappcm, PCM thermal solutions and counting. But very recently, the potential of PCMs is being explored for high-end applications such as drug-delivery, thermal barcoding, solar-powered refrigerators etc.

6.4.1. Reduction of Cooling Costs in Warehouse-Scale Computers

Due to increasing computing density of computing infrastructure, a significant portion of the initial capital expenditures and recurring operating expenditures are dedicated to cooling [196]. To prevent high server failure, the cooling system must be able to handle the peak demand of the data centre. Further, the cooling system also may become insufficient as servers undergo upgradation or replacement. To mitigate these challenges, the PCMs can be used to temporarily store the heat generated during peak load and release the heat when there is excess cooling capacity. The advantages of this approach may not be immediately felt as heat is stored only temporarily then released at a later time. However, the understanding gained from this work is that the thermal behaviour of the data centre can be designed so that the heat is released only when it is economically feasible.

6.4.2. Solar Power Refrigerators

Dulas, a Welsh renewable energy technology company, is replacing batteries by using PCM as an essential component of solar-powered direct-drive refrigerators for off-grid vaccine storage in developing countries [197]. As battery-based fridges rely on traditional evaporators to stay cool, there can be parts of the fridge where the temperature is less than 0°C, so there is a risk of freezing a vaccine resulting in its degradation. In order to improve vaccination rates, the whole fridge must be maintained at a constant temperature between 2 and 7°C. The problem is solved by having a lining with a paraffin wax-based PCM that freezes at 5°C. At night, in the absence of solar power, the PCM melts and in the process absorbs heat, without any rise in temperature, until it has all turned into liquid. This makes sure the vaccines stay within their optimum temperature range. Another advantage of the PCM design is that the thermal storage medium can perform over 10,000 cycles with minimal degradation.

6.4.3. Thermal Packaging

PCMs are also used in thermal packaging industry to maintain a temperature-sensitive product within the manufacturer's prescribed temperature range during transport [198]. As the PCM undergoing phase change temperature within the required temperature range changes its phase (for example, from solid to liquid), it effectively extends the duration of temperature control by cooling the product via latent heat. The use of electricity-free, phase change material packets in the form of an incubator has been approached to isolate bacteria for detection of typhoidal fever [199]. Such a low-cost system can be easily applied in remote areas where there is inaccessibility of sophisticated lab equipment and trained personnel.

6.4.4. Temperature-Sensitive Drug Delivery

The field of nano-drug delivery systems using external stimuli such as light, magnetic field, temperature and ultrasound is constantly being enriched with many pioneering research. As shown in Figure 14, near-infrared irradiation (NIR) can effectively deliver drug by encapsulation of photothermal agents such as gold or carbon nanomaterials in thermoresponsive polymer system [200]. Biocompatible and biodegradable organic PCMs can be utilized for temperature-sensitive drug delivery, which can be particularly useful for chemotherapeutics providing pathway for localized delivery and posing minimum side-effects [7]. Ongoing research data suggests the use of PCM microcapsules as a drug carrier and allowing its release above the melting point of the PCM. Even multiple components can be released as a function of temperature.

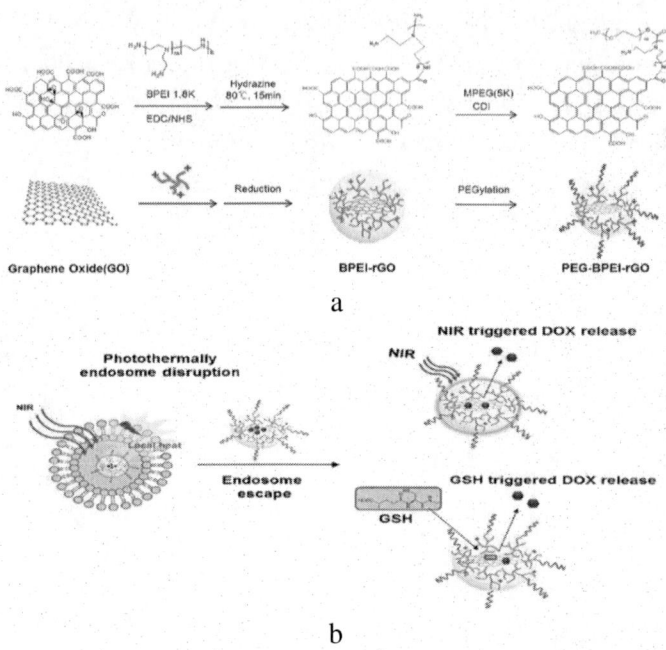

Figure 14. (a) Functionalization of reduced graphene oxide (rGO) with PEG and branched polyethylenimine (BPEI) (b) Schematic for PEG-BPEI-rGO composite loaded with anti-cancer drug, doxorubicin (DOX) and photothermally trigerred drug delivery by endosomal disruption using NIR. "Reprinted (adapted) with permission from ACS Nano 7, 8, 6735-6746. Copyright (2013) American Chemical Society."

6.4.5. Thermal Barcoding

Due to the counterfeiting associated with conventional barcodes, covert barcodes are needed to be developed for tracing documents, tracking objects and even identifying criminals or terrorists [201]. Phase change nanoparticles and their eutectic mixtures offer high loading capacity with sharp melting peaks, providing a robust system for forensic investigations. The detection of the solid-liquid phase change can be identified using infrared thermal imaging technique (Figure 15) instead of DSC to provide a non-contact, sensitive technique for decoding thermal barcodes [8].

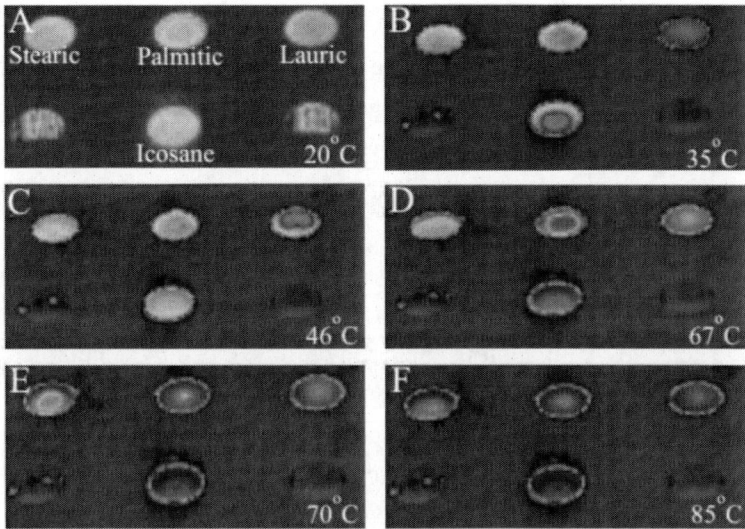

Figure 15. Infrared imaging of four different PCMs acquired over time indicating different emissivity of each material forming a code depending upon the difference in melting temperature. "Reprinted (adapted) with permission from J. Phys. Chem. C 120, 38, 22110-22114. Copyright (2016) American Chemical Society."

6.4.6. Biometric Identification

Fluorescent, thermoresponsive sensor system is developed by impregnating common filter paper with molecular rotor and phase change material by absorption in the molten state [9].

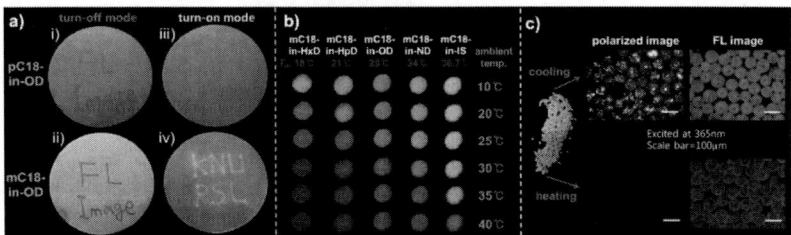

Figure 16. (a) Fluorescence imaging of characters on PCM incorporated filter paper on cooling ((i) and (ii)) and heating ((iii) and (iv)), (b) Thermoresponsive senor deposited on PCM impregnated filter paper with different melting temperatures, used as an array system and (c) polarized and corresponding flourscent microscopy images of the PCM microcapsules. "Reprinted (adapted) with permission from ACS Appl. Mater. Interfaces 7, 26, 14485-14492. Copyright (2015) American Chemical Society."

This hybrid material has the ability to respond to fluorescence change reversibly. (Figure 16) A critical change in fluorescence intensity is observed at melting and crystallization temperatures of PCM and can be effectively used for biometric identification.

6.4.7. Anti-Icing Coating

Ice accumulation on aircrafts or runways can lead to fatal accidents and the currently available mechanical and chemicals methods are labour and time-intensive. An ingenious method was developed by Bhamidipati of E-paint company to prevent accumulation of ice using bio-inspired anti-icing coatings comprising silicone based PCMs, preferably a hydrophobic resin [202]. With ice-formation, phase change occurs with PCM expanding and substrate contracting. The latent heat released from freezing of water is passed on to the PCM near the surface thereby experiencing a local shear stress due to volume change and ultimately, resulting in disruption of ice-surface bond.

6.4.8. Thermal Peak Management Using Organic Phase Change Materials

A composite coating for electronic components on printed circuit boards or electronic assemblies is expected to buffer a certain amount of thermal energy, dissipated from a device [203]. To prevent damage to the instrument during temperature peaks in electronic components, a phase change material can be used. The phase change material can be coated directly on the chip package or the PCB using different mechanical retaining jigs. Sugar alcohols as PCM and epoxy or acrylate as resin matrix, together with metal nanoparticles or CNT as thermal conductivity enhancer are used for making the coating.

6.4.9. Paraffin Wax in 3D Printing

For potential applications in winter sport equipment, 3D printable blends with thermal energy storage (TES) capabilities can be developed by incorporating paraffin in thermoplastic polyurethane (TPU) [204]. Thoroughly mixed encapsulated paraffin and TPU granules are first extruded and then the filaments are used for 3D printing using the set up: Layer height 0.20 mm, Nozzle temperature 240 °C, Bed temperature 40 °C, Deposition rate 40 mm s^{-1}. Homogeneous distribution of the capsules in the polymer matrix and a good adhesion between the layers in the 3D printed parts can be achieved by this technique. Reasonable energy storage/release capability is obtained in the 3D printed parts, with melting enthalpy values reaching up to 70 J g^{-1}. The elastic modulus increases with the microcapsule content and higher elastic modulus of melamine formaldehyde shell of the capsules, as compared to TPU matrix, generate stiffening of the matrix. The hard microcapsules also increase creep resistance and Shore A hardness of the material.

7. ENVIRONMENTAL IMPACT OF PCMS

In the literature, numerous PEG based SSPCMs have been examined with respect to their chemical, thermal and morphological characteristics. One important aspect to be taken into consideration during PCM preparation is the need to develop environment and user-friendly materials with biodegradable and non-toxic nature.

PEG, a synthetic biomaterial is innocuous and widely used in the production of cosmetics and pharmaceuticals. Even high molecular weight PEGs are susceptible to biodegradation [205]. Due to the abundant availability, renewability, low-cost and biodegradability of the natural polymers, many researchers have integrated soft segment PEG with natural polymers as hard segments and achieved solid-solid phase transformation with satisfactory thermal properties. Composites of PEG with natural polymers including cellulose [164], cellulose acetate [37], carboxymethyl cellulose [70], agarose, chitosan [71], starch [107], sugars (glucose, fructose and lactose) [108] have been prepared by using both physical blending and chemical grafting methods. β-cyclodextrin with α-glucopyranose units, widely used for biomedical applications, is cross-linked with PEG in the presence of MDI to prepare SSPCM [145].

Similarly, trihydroxy surfactants based on natural fatty acids, Span 80 and tween 80 are used as cross-linking agents in the solventless preparation of PEG based polyurethanes [146]. Thermoplastic polyester, PHBV produced naturally by bacteria can be copolymerized with PEG diacrylate to form environment friendly copolymer PCM [182]. The utilization of cotton waste in the development of thermoregulating fabric by grafting of PEG will aid in industrial waste management [165].

Solvent-free synthesis of PEG based SSPCM, with vegetable oil as hard segment linked by urethane bonds, can be carried out leading to advanced energy storage material [151]. It is believed that the use of biodegradable starting materials in the preparation of PCM can benefit the biodegradability of the prepared SSPCM. The microwave route can be adopted for green synthesis of PEG-CDA composites with minimum use of solvent [69]. The energy and time efficient technique follows the principles of green chemistry. All these TES materials are biocompatible, biodegradable and environmentally safe macromolecules which should be utilized in the development of advanced SSPCMs.

CONCLUSION

A wide array of polymers have been discussed which can be prepared by using both physical and chemical methods, capable of undergoing thermal energy storage. Among the physical methods, blending has been most commonly utilized to prepare solid-solid PCMs whereas urethane modification has been most commonly applied to prepare chemically bonded PCMs. Several chemical methods of modification have been utilized to prepare polymers for thermal energy storage including grafting, crosslinking, copolymerization and hyperbranching. The use of several modification techniques listed offer a wide array of transition temperatures and enthalpies which can accommodate user's requirements.

REFERENCES

[1] Farid, MM, Khudhair, AM, Razack, SAK, Al-Hallaj, S. 2004. "A Review on Phase Change Energy Storage: Materials and Applications." *Energy Convers. Manag.* 45 (9–10): 1597–615.

[2] Pielichowska, K, Pielichowski, K. 2014. "Phase Change Materials for Thermal Energy Storage." *Prog. Mater. Sci.* 65: 67–123.

[3] Odian, G. 2004. *Principles of Polymerization (Fourth Edition).* New York, USA: Wiley-Interscience.

[4] Sharma, A, Tyagi, VV, Chen, CR, Buddhi, D. 2009. "Review on Thermal Energy Storage with Phase Change Materials and Applications." *Renew. Sustain. Energy Rev.* 13 (2): 318–45.

[5] Mehling, H, Cabeza, LF. 2008. "Solid-Liquid Phase Change Materials." In *Heat and Cold Storage with PCM: An up to Date Introduction into Basics and Applications,* 11–55. Heidelberg, Berlin: Springer.

[6] Zalba, B, Marín, JM, Cabeza, LF, Mehling, H. 2003. "Review on Thermal Energy Storage with Phase Change: Materials, Heat Transfer Analysis and Applications." *Appl. Therm. Eng.* 23 (3): 251–83.

[7] Hyun, DC, Levinson, NS, Jeong, U, Xia, Y. 2014. "Emerging Applications of Phase-Change Materials (PCMs): Teaching an Old Dog New Tricks." *Angew. Chem. Int. Ed.* 53 (15): 3780–95.

[8] Hou, S, Zheng, W, Duong, B, Su, M. 2016. "All-Optical Decoder for Rapid and Noncontact Readout of Thermal Barcodes." *J. Phys. Chem. C* 120 (38): 22110–14.

[9] Jin, YJ, Dogra, R, Cheong, IW, Kwak, G. 2015. "Fluorescent Molecular Rotor-in-Paraffin Waxes for Thermometry and Biometric Identification." *ACS Appl. Mater. Interfaces* 7 (26): 14485–92.

[10] Kenisarin, MM, Kenisarina, KM. 2012. "Form-Stable Phase Change Materials for Thermal Energy Storage. Renew." *Sustain. Energy Rev.* 16 (4): 1999–2040.

[11] Meng, Q, Hu, J. 2008. "A Poly(Ethylene Glycol)-Based Smart Phase Change Material." *Sol. Energy Mater. Sol. Cells* 92 (10): 1260–68.

[12] Hussein, HA, Abed, AH, Abdulmunem, AR. 2016. "An Experimental Investigation of Using Aluminum Foam Matrix Integrated with Paraffin Wax as a Thermal Storage Material in a Solar Heater." In *Proceeding of the 2nd Sustainable & Renewable Energy Conference.* Baghdad-Iraq. 6-7 Dec, 2017.

[13] Dorigato, A, Ciampolillo, MV, Cataldi, A, Bersani, M, Pegoretti, A. 2017. "Polyethylene Wax/EPDM Blends as Shape-Stabilized Phase Change Materials for Thermal Energy Storage." *Rubber Chem. Technol.* 90 (3): 575–84.

[14] Fredi, G, Dorigato, A, Pegoretti, A. 2018. "Multifunctional Glass Fiber/Polyamide Composites with Thermal Energy Storage/Release Capability." *Express Polym. Lett.* 12 (4): 349–64.

[15] Fredi, G, Dorigato, A, Fambri, L, Pegoretti, A. 2017. "Wax Confinement with Carbon Nanotubes for Phase Changing Epoxy Blends." *Polymers* 9 (12): 405.

[16] Yang, J, Qi, GQ, Liu, Y, Bao, RY, Liu, ZY, Yang, W, Xie, BH, Yang, MB. 2016. "Hybrid Graphene Aerogels/Phase Change Material Composites: Thermal Conductivity, Shape-Stabilization and Light-to-Thermal Energy Storage." *Carbon* 100: 693–702.

[17] Fredi, G, Andrea, D, Luca, F, Alessandro, P. 2018. "Multifunctional Epoxy/Carbon Fiber Laminates for Thermal Energy Storage and Release." *Compos. Sci. Technol.* 158: 101–11.

[18] Liu, Z, Chen, Z, Yu, F. 2018. "Microencapsulated Phase Change Material Modified by Graphene Oxide with Different Degrees of Oxidation for Solar Energy Storage." *Sol. Energy Mater. Sol. Cells* 174: 453–59.

[19] Yang, J, Zhang, E, Li, X, Zhang, Y, Qu, J, Yu, ZZ. 2016. "Cellulose/Graphene Aerogel Supported Phase Change Composites with High Thermal Conductivity and Good Shape Stability for Thermal Energy Storage." *Carbon* 98: 50–7.

[20] Zheng, L, Zhang, W, Liang, F. 2017. "A Review about Phase Change Material Cold Storage System Applied to Solar-Powered Air-Conditioning System." *Adv. Mech. Eng.* 9 (6): 1-20.

[21] Lei, J, Yang, J, Yang, EH. 2016. "Energy Performance of Building Envelopes Integrated with Phase Change Materials for Cooling Load Reduction in Tropical Singapore." *Appl. Energy* 162: 207–17.

[22] Shin, Y, Yoo, DII, Son, K. 2005. "Development of Thermoregulating Textile Materials with Microencapsulated Phase Change Materials (PCM). IV. Performance Properties and Hand of Fabrics Treated with PCM Microcapsules." *J. Appl. Polym. Sci.* 97 (3): 910–15.

[23] Zhang, H, Xu, Q, Zhao, Z, Zhang, J, Sun, Y, Sun, L, Xu, F, Sawada, Y. 2012. "Preparation and Thermal Performance of Gypsum Boards Incorporated with Microencapsulated Phase Change Materials for Thermal Regulation." *Sol. Energy Mater. Sol. Cells* 102: 93–102.

[24] Song, Q, Li, Y, Xing, J, Hu, JY, Marcus, Y. 2007. "Thermal Stability of Composite Phase Change Material Microcapsules Incorporated with Silver Nano-Particles." *Polymer* 48 (11): 3317–23.

[25] Yin, D, Liu H, Ma L, Zhang Q. 2015. "Fabrication and Performance of Microencapsulated Phase Change Materials with Hybrid Shell by in Situ Polymerization in Pickering Emulsion." *Polym. Adv. Technol.* 26 (6): 613–19.

[26] Jiang, X, Luo, R, Peng, F, Fang, Y, Akiyama, T, Wang, S. 2015. "Synthesis, Characterization and Thermal Properties of Paraffin Microcapsules Modified with Nano-Al_2O_3." *Appl. Energy* 137: 731–37.

[27] Do, T, Ko, YG, Chun, Y, Choi, US. 2015. "Encapsulation of Phase Change Material with Water-Absorbable Shell for Thermal Energy Storage." *ACS Sustain. Chem. Eng.* 3 (11): 2874–81.

[28] Sun, Y, Wang, R, Liu, X, Li, M, Yang, H, Li, B. 2017. "Improvements in the Thermal Conductivity and Mechanical Properties of Phase-Change Microcapsules with Oxygen-Plasma-Modified Multiwalled Carbon Nanotubes." *J. Appl. Polym. Sci.* 134 (44): 45269.

[29] Chen, L, Zou, R, Xia, W, Liu, Z, Shang, Y, Zhu, J, Wang, Y, Lin, J, Xia, D, Cao, A. 2012. "Electro- and Photodriven Phase Change

Composites Based on Wax-Infiltrated Carbon Nanotube Sponges." *ACS Nano* 6 (12): 10884–92.

[30] Wang, Y, Tang, B, Zhang, S. 2012. "Novel Organic Solar Thermal Energy Storage Materials: Efficient Visible Light-Driven Reversible Solid–liquid Phase Transition." *J. Mater. Chem.* 22 (35): 18145-50.

[31] Wang, Y, Bingtao, T, Shufen, Z. 2012. "Visible Light-Driven Organic Form-Stable Phase Change Materials for Solar Energy Storage." *RSC Adv.* 2 (14): 5964-67.

[32] Qi, G, Yang, J, Bao, R, Xia, D, Cao, M, Yang, W, Yang, M, Wei, D. 2017. "Hierarchical Graphene Foam-Based Phase Change Materials with Enhanced Thermal Conductivity and Shape Stability for Efficient Solar-to-Thermal Energy Conversion and Storage." *Nano Res.* 10 (3): 802–13.

[33] Li, M, Guo, Q, Nutt, S. 2017. "Carbon Nanotube/Paraffin/ Montmorillonite Composite Phase Change Material for Thermal Energy Storage." *Sol. Energy* 146: 1–7.

[34] Sundararajan, S, Samui, AB, Kulkarni, PS. 2017. "Synthesis and Characterization of Poly(Ethylene Glycol) (PEG) Based Hyperbranched Polyurethanes as Thermal Energy Storage Materials." *Thermochim. Acta* 650: 114–22.

[35] McCann, JT, Marquez, M, Xia, YN. 2006. "Melt coaxial electrospinning: a versatile method for the encapsulation of solid materials and fabrication of phase change nanofibers." *Nano Lett.* 6 (12): 2868–72.

[36] Bergshoef, MM, Vancso, GJ. 1999. "Transparent nanocomposites with ultrathin, electrospun nylon-4,6 fiber reinforcement." *Adv Mater* 11 (16): 1362–65.

[37] Chen, CZ, Wang, L, Huang, Y. 2007. "Electrospinning of thermoregulating ultrafine fibers based on polyethylene glycol/cellulose acetate composite." *Polymer* 48 (18): 5202–07.

[38] Chen, CZ, Wang, LG, Huang, Y. 2008. "A novel shape-stabilized PCM: electrospun ultrafine fibers based on lauric acid/polyethylene terephthalate composite." *Mater Lett* 62 (20): 3515–17.

[39] Mondal, S. 2008. "Phase change materials for smart textiles – An overview." *Appl. Therm. Eng.* 28: 1536–50.

[40] Pérez-Masiá, R, López-Rubio, A, Lagarón, JM. 2013. "Development of zein-based heat management structures for smart food packaging." *Food Hydrocolloids* 30 (1): 182–91.

[41] Li, F, Zhao, Y, Wang, S, Han, D, Jiang, L, Song, Y. 2009. "Thermochromic core-shell nanofibers fabricated by melt coaxial electrospinning." *J Appl Polym Sci* 112 (1): 269–74.

[42] Shi, Q, Liu, Z, Jin, X, Shen, Y, Liu, Y. 2015. "Electrospun fibers based on polyvinyl pyrrolidone/Eu-polyethylene glycol as phase change luminescence materials." *Mater Lett.* 147:113–15.

[43] Abidian, MR, Martin, DC, 2009. "Multifunctional Nanobiomaterials for Neural Interfaces." *Adv. Funct. Mater.* 19: 573-85.

[44] Hassler, C, Boretius, T, Stieglitz, T. 2011. "Polymers for neural implants." *J. Polym. Sci., Part B: Polym. Phys.* 49: 18-33.

[45] Chen, C, Wang, L, Huang, Y. 2008. "Morphology and thermal properties of electrospun fatty acids/polyethylene terephthalate composite fibers as novel form-stable phase change materials." *Sol. Energy Mater. Sol. Cells* 92: 1382–87.

[46] Chen, C, Liu, S, Liu, W, Zhao, Y, Lu, Y. 2012. "Synthesis of novel solid–liquid phase change materials and electrospinning of ultrafine phase change fibers." *Sol. Energy Mater. Sol. Cells* 96: 202–09.

[47] Chen, C, Wang, L, Huang, Y. 2011. "Electrospun phase change fibers based on polyethylene glycol/cellulose acetate blends." *Appl. Energy* 88: 3133–39.

[48] Chen, C, Wang, L, Huang, Y. 2009. "Crosslinking of the electrospun polyethylene glycol/cellulose acetate composite fibers as shape-stabilized phase change materials." *Mater. Lett.* 63: 569–71

[49] Chen, H, Ma, Q, Wang, S, Liu, H, Wang, K. 2016. *Morphology, compatibility, physical and thermoregulated properties of the electrospinning polyamide 6 and polyethylene glycol blended nanofibers.* 45 (6): 1490-1503,

[50] Zhao, L, Luo, J, Li, Y, Wang, H, Song, G, Tang, G. 2017. "Emulsion-electrospinning n-octadecane/silk composite fiber as

environmental-friendly form-stable phase change materials." *J. Appl. Polym. Sci.* 134: 45538.

[51] Ke, G, Wang, X, Pei, J. 2018, "Fabrication and properties of electrospun PAN/LA-SA/TiO$_2$ composite phase change fiber." *Polymer-Plastics Technol. Eng.* 57 (10): 958-64.

[52] Zhang, J, Yu, J, Cai, Y, Lv, P, Zhou, H, Wei, Q. 2019. "Fabrication of Form-Stable Phase Change Materials Based on Mechanically Flexible SiO$_2$ Nanofibrous Mats for Thermal Energy Storage/Retrieval." *J. Nanosci. Nanotechnol.* 19: 5562–71.

[53] Wan, Y, Zhou, P, Liu, Y, Chen, H. 2016. "Novel wearable polyacrylonitrile/phase-change material sheath/core nano-fibers fabricated by coaxial electro-spinning." *RSC Adv.* 6: 21204-09.

[54] Ranodhi, WM, Udangawa, N, Willard, CF, Mancinelli, C, Chapman, C, Linhardt, RJ, Simmons, TJ. 2019. "Coconut oil-cellulose beaded microfibers by coaxial electrospinning: An eco-model system to study thermoregulation of confined phase change materials." *Cellulose* 26:1855–1868.

[55] Zheng, Y, Cai, C, Zhang, F, Monty, J, Linhardt, RJ, Simmons, TJ. 2016. "Can natural fibers be a silver bullet? Antibacterial cellulose fibers through the covalent bonding of silver nanoparticles to electrospun fibers." *Nanotechnology* 27: 055102.

[56] Xi, P, Zhao, T, Xia, L, Shu, D, Ma, M, Cheng, B. 2017. "Fabrication and characterization of dual-functional ultrafine composite fibers with phase-change energy storage and luminescence properties." *Sci. Rep.* 7: 40390.

[57] Shi, Q, Liu, Z, Jin, X, Shen, Y, Liu, Y. 2015. "Electrospun fibers based on polyvinyl pyrrolidone/Eu-polyethylene Glycol as phase change luminescence materials." *Mater. Lett.* 147: 113–15.

[58] Xia, X, Zhang, Y, Li, Q, Li, C. 2014. "Preparation and Characterization of thermochromic phase change nanofibers/ woven composite material." *Adv. Mater. Res.* 1048: 427-31.

[59] Lu, Y, Xiao, X, Fu, J, Huan, C, Qi, S, Zhan, Y, Zhu, Y, Xu, G. 2019. "Novel Smart Textile with Phase Change Materials

Encapsulated Core-Sheath Structure Fabricated by Coaxial Electrospinning." *Chem. Eng. J.* 355: 532-39.

[60] Cai, Y, Liu, M, Song, X, Zhang, J, Wei, Q, Zhang, L. 2015. "A form-stable phase change material made with a cellulose acetate nanofibrous mat from bicomponent electrospinning and incorporated capric–myristic–stearic acid ternary eutectic mixture for thermal energy storage/retrieval." *RSC Adv.* 5: 84245–51

[61] Baştürk, E, Deniz, DY, Kahraman, MV, 2019. "Form-stable electrospun nanofibrous mats as a potential phase change material." *J. Macromol. Sci. Part A: Pure and Appl. Chem.* 56: 708–16.

[62] Chalco-Sandoval, W, Fabra, MJ, López-Rubio, A, Lagaron, JM, 2014. "Electrospun heat management polymeric materials of interest in food refrigeration and packaging." *J. Appl. Polym. Sci.* 131: 40661.

[63] Chalco-Sandoval, W, Fabra, MJ, López-Rubio, A, Lagaron, JM, 2017. "Use of phase change materials to develop electrospun coatings of interest in food packaging applications." *J. Food Eng.* 192: 122-28.

[64] Xue, J, Zhu, C, Li, J, Li, H, Xia, Y. 2018. "Integration of Phase-Change Materials with Electrospun Fibers for Promoting Neurite Outgrowth under Controlled Release." *Adv. Funct. Mater.* 28: 1705563.

[65] Molefi, JA. 2008. *Investigation of Phase Change Conducting Materials Prepared From Polyethylenes*, Paraffin Waxes and Copper. PhD thesis, University of the Free State.

[66] Liang, XH, Guo, YQ, Gu, LZ, Ding, EY. 1995. "Crystalline-Amorphous Phase Transition of Poly(Ethylene Glycol)/Cellulose Blend." *Macromolecules* 28: 6551–55.

[67] Ding, EY, Jiang, Y, Li, GK. 2001. "Comparative Studies of the Structures and Transition Characteristics of Cellulose Diacetate Modified with Polyethylene Glycol Prepared by Chemical Bonding and Physical Blending Methods." *J. Macromol. Sci., Part B* 40: 1053–1068.

[68] Guo, Y, Tong, Z, Chen, M, Liang, X. 2003. "Solution miscibility and phase-change behavior of a polyethylene glycol-diacetate cellulose composite." *J. Appl. Polym. Sci.* 88: 652–58.
[69] Sundararajan, S, Samui, AB, Kulkarni, PS. 2017. "Shape-Stabilized Poly(Ethylene Glycol) (PEG)-Cellulose Acetate Blend Preparation with Superior PEG Loading via Microwave-Assisted Blending." *Sol. Energy* 144: 32–9.
[70] Pielichowska, K, Pielichowski, K. 2011. "Biodegradable PEO/Cellulose-Based Solid-Solid Phase Change Materials." *Polym. Adv. Technol.* 22 (12): 1633–41.
[71] Şentürk, SB, Kahraman, D, Alkan, C, Gökçe, I. 2011. "Biodegradable PEG/Cellulose, PEG/Agarose and PEG/Chitosan Blends as Shape Stabilized Phase Change Materials for Latent Heat Energy Storage." *Carbohydrate Polym.* 84 (1): 141–44.
[72] Cao, L, Tang, Y, Fang, G. 2015. "Preparation and Properties of Shape-Stabilized Phase Change Materials Based on Fatty Acid Eutectics and Cellulose Composites for Thermal Energy Storage." *Energy* 80: 98–103.
[73] Alkan, C, Sari A, Uzun O. 2006. "Poly(Ethylene Glycol)/Acrylic Polymer Blends for Latent Heat Thermal Energy Storage." *AIChE Journal* 52 (9): 3310–14.
[74] Alkan, C, Günther, E, Hiebler, S, Himpel, M. 2012. "Complexing Blends of Polyacrylic Acid-Polyethylene Glycol and Poly(Ethylene-Co-Acrylic Acid)-Polyethylene Glycol as Shape Stabilized Phase Change Materials." *Energy Conv. Manag.* 64: 364–70.
[75] Wang, L, Meng, D. 2010. "Fatty Acid Eutectic/Polymethyl Methacrylate Composite as Form-Stable Phase Change Material for Thermal Energy Storage." *Appl. Energy* 87 (8): 2660–65.
[76] Zhang, L, Zhu, J, Zhou, W, Wang, J, Wang, Y. 2011. "Characterization of Polymethyl Methacrylate/Polyethylene Glycol/Aluminum Nitride Composite as Form-Stable Phase Change Material Prepared by in Situ Polymerization Method." *Thermochim. Acta* 524 (1–2): 128–34.

[77] Zhang, L, Zhu, J, Zhou, W, Wang, J, Wang, Y. 2012. "Thermal and Electrical Conductivity Enhancement of Graphite Nanoplatelets on Form-Stable Polyethylene Glycol/Polymethyl Methacrylate Composite Phase Change Materials." *Energy* 39 (1): 294–302.

[78] Chen, C, Liu, W, Wang, Z, Peng, K, Pan, W, Xie, Q. 2015. "Novel form stable phase change materials based on the composites of polyethylene glycol/polymeric solid-solid phase change material." *Sol. Energy Mater. Sol. Cells.* 134: 80–88.

[79] Chen, K, Yu, X, Tian, C, Wang, J. 2014. "Preparation and Characterization of Form-Stable Paraffin/Polyurethane Composites as Phase Change Materials for Thermal Energy Storage." *Energy Convers. Manag.* 77: 13–21.

[80] Yilgor, I, Yilgor, E. 2007. "Structure-Morphology-Property Behavior of Segmented Thermoplastic Polyurethanes and Polyureas Prepared without Chain Extenders." *Polymer Rev.* 47 (4): 487–510.

[81] Pielichowska, K, Nowak, M, Szatkowski, P, Macherzyńska, B. 2016. "The Influence of Chain Extender on Properties of Polyurethane-Based Phase Change Materials Modified with Graphene." *Appl. Energy* 162: 1024–33.

[82] Kong, W, Fu, X, Yuan, Y, Liu, Z, Lei, J. 2017. "Preparation and Thermal Properties of Crosslinked Polyurethane/Lauric Acid Composites as Novel Form Stable Phase Change Materials with a Low Degree of Supercooling." *RSC Adv.* 7 (47): 29554–62.

[83] Zhang, Y, Wang, L, Tang, B, Lu, R, Zhang, S. 2016. "Form-Stable Phase Change Materials with High Phase Change Enthalpy from the Composite of Paraffin and Cross-Linking Phase Change Structure." *Appl. Energy* 184: 241–46.

[84] Tang, B, Wang, L, Xu, Y, Xiu, J, Zhang, S. 2016. "Hexadecanol/Phase Change Polyurethane Composite as Form-Stable Phase Change Material for Thermal Energy Storage." *Sol. Energy Mater. Sol. Cells* 144: 1–6.

[85] Feng, L, Zhao, W, Zheng, J, Frisco, S, Song, P, Li, X. 2011. "The Shape-Stabilized Phase Change Materials Composed of

Polyethylene Glycol and Various Mesoporous Matrices (AC, SBA-15 and MCM-41)." *Sol. Energy Mater. Sol. Cells* 95 (12): 3550–56.

[86] Zhang, L, Shi, H, Li, W, Han, X, Zhang, X. 2014. "Thermal Performance and Crystallization Behavior of Poly(Ethylene Glycol) Hexadecyl Ether in Confined Environment." *Polym. Int.* 63 (6): 982–88.

[87] Wang, C, Feng, L, Li, W, Zheng, J, Tian, W, Li, X. 2012. "Shape-Stabilized Phase Change Materials Based on Polyethylene Glycol/Porous Carbon Composite: The Influence of the Pore Structure of the Carbon Materials." *Sol. Energy Mater. Sol. Cells* 105: 21–26.

[88] Feng, L, Wang, C, Song, P, Wang, H, Zhang, X. 2015. "The Form-Stable Phase Change Materials Based on Polyethylene Glycol and Functionalized Carbon Nanotubes for Heat Storage." *Appl. Therm. Eng.* 90: 952–56.

[89] Tang, J, Yang, M, Dong, W, Yang, M, Zhang, H, Fan, S, Wang, J, Tan, L, Wang, G. 2016. "Highly Porous Carbons Derived from MOFs for Shape-Stabilized Phase Change Materials with High Storage Capacity and Thermal Conductivity." *RSC Adv.* 6 (46): 40106–14.

[90] Tan, B, Huang, Z, Yin, Z, Min, X, Liu, Y, Wu, X, Fang, M. 2016. "Preparation and Thermal Properties of Shape-Stabilized Composite Phase Change Materials Based on Polyethylene Glycol and Porous Carbon Prepared from Potato." *RSC Adv.* 6 (19): 15821–30.

[91] Wang, W, Xiaoxi, Y, Yutang, F, Jing, D, Jinyue, Y. 2009b. "Preparation and Thermal Properties of Polyethylene Glycol/Expanded Graphite Blends for Energy Storage." *Appl. Energy* 86 (9): 1479–83.

[92] Jiang, Y, Wang, Z, Shang, M, Zhang, Z, Zhang, S. 2015. "Heat Collection and Supply of Interconnected Netlike Graphene/Polyethyleneglycol Composites for Thermoelectric Devices." *Nanoscale* 7 (25): 10950–53.

[93] Li, H, Jiang, M, Li, Q, Li, D, Chen, Z, Hu, W, Huang, J, Xu, X, Dong, L, Xie, H, Xiong, C. 2013. "Aqueous Preparation of

Polyethylene Glycol/Sulfonated Graphene Phase Change Composite with Enhanced Thermal Performance." *Energy Conver. Manag.* 75: 482–487.

[94] Wang, C, Feng, L, Yang, H, Xin, G, Li, W, Zheng, J, Tian, W, Li, X. 2012. "Graphene Oxide Stabilized Polyethylene Glycol for Heat Storage." *Phys. Chem. Chem. Phys.* 14 (38): 13233–38.

[95] Qi, GQ, Liang, CL, Bao, RY, Liu, ZY, Yang, W, Xie, BH, Yang, MB. 2014. "Polyethylene Glycol Based Shape-Stabilized Phase Change Material for Thermal Energy Storage with Ultra-Low Content of Graphene Oxide." *Sol. Energy Mater. Sol. Cells* 123: 171–77.

[96] Xiong, W, Chen, Y, Hao, M, Zhang, L, Mei, T, Wang, J, Li, J, Wang, X. 2015. "Facile Synthesis of PEG Based Shape-Stabilized Phase Change Materials and Their Photo-Thermal Energy Conversion." *Appl. Therm. Eng.* 91: 630–37.

[97] Wang, C, Wang, W, Xin, G, Li, G, Zheng, J, Tian, W, Li, X. 2016. "Phase Change Behaviors of PEG on Modified Graphene Oxide Mediated by Surface Functional Groups." *Eur. Polym. J.* 74: 43–50.

[98] Akhiani, AR, Mehrali, M, Latibari, ST, Mehrali, M, Mahlia, TMI, Sadeghinezhad, E, Metselaar, HSC. 2015. "One-Step Preparation of Form-Stable Phase Change Material through Self-Assembly of Fatty Acid and Graphene." *J. Phys. Chem. C* 119 (40): 22787–96.

[99] Qi, GQ, Yang, J, Bao, RY, Liu, ZY, Yang, W, Xie, BH, Yang, MB. 2015. "Enhanced Comprehensive Performance of Polyethylene Glycol Based Phase Change Material with Hybrid Graphene Nanomaterials for Thermal Energy Storage." *Carbon* 88: 196–205.

[100] Fang, Y, Kang, H, Wang, W, Liu, H, Gao, X. 2010. "Study on Polyethylene Glycol/Epoxy Resin Composite as a Form-Stable Phase Change Material." *Energy Convers. Manag.* 51 (12): 2757–61.

[101] Mu, S, Guo, J, Yu, Y, An, Q, Zhang, S, Wang, D, Chen, S, Huang, X, Li, S. 2016. "Synthesis and Thermal Properties of Cross-Linked Poly(Acrylonitrile-Co-Itaconate)/Polyethylene Glycol as Novel

Form-Stable Change Material." *Energy Convers. Manag.* 110: 176–83.

[102] Cai, Y, Wei, Q, Huang, F, Gao, W. 2008. "Preparation and Properties Studies of Halogen-Free Flame Retardant Form-Stable Phase Change Materials Based on Paraffin/High Density Polyethylene Composites." *Appl. Energy* 85 (8): 765–75.

[103] Hong, H, Pan, Y, Sun, H, Zhu, Z, Ma, C, Wang, B, Liang, W, Yang, B, Li, A. 2018. "Superwetting Polypropylene Aerogel Supported Form-Stable Phase Change Materials with Extremely High Organics Loading and Enhanced Thermal Conductivity." *Sol. Energy Mater. Sol. Cells* 174: 307–13.

[104] Zeng, JL, Zhang, J, Liu, YY, Cao, ZX, Zhang, ZH, Xu, F, Sun, LX. 2008. "Polyaniline/1-Tetradecanol Composites." *J. Therm. Anal. Calorim.* 91 (2): 455–61.

[105] Zeng, JL, Zhu, FR, Yu, SB, Xiao, ZL, Yan, WP, Zheng, SH, Zhang, L, Sun, LX, Cao, Z. 2013. "Myristic Acid/Polyaniline Composites as Form Stable Phase Change Materials for Thermal Energy Storage." *Sol. Energy Mater. Sol. Cells* 114: 136–40.

[106] Zeng, JL, Zheng, SH, Yu, SB, Zhu, FR, Gan, J, Zhu, L, Xiao, ZL, Zhu, XY, Zhu, Z, Sun, LX, Cao, Z. 2014. "Preparation and Thermal Properties of Palmitic Acid/Polyaniline/Exfoliated Graphite Nanoplatelets Form-Stable Phase Change Materials." *Appl. Energy* 115: 603–09.

[107] Pielichowska, K, Pielichowski, K. 2010. "Novel Biodegradable Form Stable Phase Change Materials: Blends of Poly(Ethylene Oxide) and Gelatinized Potato Starch." *J. Appl. Polym. Sci.* 21 (7): 1725–31.

[108] Alkan, C, Günther, E, Hiebler, S, Ensari, OF, Kahraman, D. 2012. "Polyethylene Glycol-Sugar Composites as Shape Stabilized Phase Change Materials for Thermal Energy Storage." *Polym. Compos.* 33 (10): 1728–36.

[109] Wang, W, Yang, X, Fang, Y, Ding, J. 2009. "Preparation and Performance of Form-Stable Polyethylene Glycol/Silicon Dioxide

Composites as Solid-Liquid Phase Change Materials." *Appl. Energy* 86 (2): 170–74.

[110] Yang, H, Feng, L, Wang, C, Zhao, W, Li, X. 2012. "Confinement Effect of SiO_2 Framework on Phase Change of PEG in Shape-Stabilized PEG/SiO_2 Composites." *Eur. Polym. J.* 48 (4): 803–10.

[111] Tang, B, Cui, J, Wang, Y, Jia, C, Zhang, S. 2013. "Facile Synthesis and Performances of PEG/SiO2 Composite Form-Stable Phase Change Materials." *Sol. Energy* 97: 484–92.

[112] Li, H, Jiang, M, Li, Q, Li, D, Chen, Z, Hu, W, Huang, J, Xu, X, Dong, L, Xie, H, Xiong, C. 2013. "Aqueous Preparation of Polyethylene Glycol/Sulfonated Graphene Phase Change Composite with Enhanced Thermal Performance." *Energy Conver. Manag.* 75: 482–87.

[113] He, L, Li, J, Zhou, C, Zhu, H, Cao, X, Tang, B. 2014. "Phase Change Characteristics of Shape-Stabilized PEG/SiO2 Composites Using Calcium Chloride-Assisted and Temperature-Assisted Sol Gel Methods." *Sol. Energy* 103: 448–55.

[114] Qian, T, Li, J, Ma, H, Yang, J. 2015. "The Preparation of a Green Shape-Stabilized Composite Phase Change Material of Polyethylene Glycol/SiO_2 with Enhanced Thermal Performance Based on Oil Shale Ash via Temperature-Assisted Sol–gel Method." *Sol. Energy Mater. Sol. Cells* 132: 29–39.

[115] Wang, W, Yang, X, Fang, Y, Ding, J, Yan, J. 2009. "Enhanced Thermal Conductivity and Thermal Performance of Form-Stable Composite Phase Change Materials by Using β-Aluminum Nitride." *Appl. Energy* 86 (7–8): 1196–200.

[116] Tang, B, Qiu, M, Zhang, S. 2012. "Thermal Conductivity Enhancement of PEG/SiO2 Composite PCM by in Situ Cu Doping." *Sol. Energy Mater. Sol. Cells* 105: 242–48.

[117] Tang, B, Wang, Y, Qiu, M, Zhang, S. 2014. "A Full-Band Sunlight-Driven Carbon Nanotube/PEG/SiO2 Composites for Solar Energy Storage." *Sol. Energy Mater. Sol. Cells* 123: 7–12.

[118] Elgafy, A, Lafdi, K. 2005. "Effect of Carbon Nanofiber Additives on Thermal Behavior of Phase Change Materials." *Carbon* 43 (15): 3067–74.

[119] Feng, L, Zheng J, Yang H, Guo Y, Li W, Li X. 2011. "Preparation and Characterization of Polyethylene Glycol/Active Carbon Composites as Shape-Stabilized Phase Change Materials." *Sol. Energy Mater. Sol. Cells* 95 (2): 644–50.

[120] Zhang, L, Shi, H, Li, W, Han, X, Zhang, X. 2013. "Structure and Thermal Performance of Poly(Ethylene Glycol) Alkylether (Brij)/Porous Silica (MCM-41) Composites as Shape-Stabilized Phase Change Materials." *Thermochim. Acta* 570: 1–7.

[121] Min, X, Fang, M, Huang, Z, Liu, Y, Huang, Y, Wen, R, Qian, T, Wu, X. 2015. "Enhanced Thermal Properties of Novel Shape-Stabilized PEG Composite Phase Change Materials with Radial Mesoporous Silica Sphere for Thermal Energy Storage." *Sci. Rep.* 5: 12964.

[122] Mitran, RA, Berger, D, Munteanu, C, Matei, C. 2015. "Evaluation of Different Mesoporous Silica Supports for Energy Storage in Shape-Stabilized Phase Change Materials with Dual Thermal Responses." *J. Phys. Chem. C* 119 (27): 15177–84.

[123] Wang, J, Yang, M, Lu, Y, Jin, Z, Tan, L, Gao, H, Fan, S, Dong, W, Wang, G. 2016. "Surface Functionalization Engineering Driven Crystallization Behavior of Polyethylene Glycol Confined in Mesoporous Silica for Shape-Stabilized Phase Change Materials." *Nano Energy* 19: 78–87.

[124] Li, H, Fang, GY. 2010. "Experimental Investigation on the Characteristics of Polyethylene Glycol/Cement Composites as Thermal Energy Storage Materials." *Chem. Eng. Technol.* 33 (10): 1650–54.

[125] Qian, T, Jinhong, Li, Xin, Min, Yong, Deng, Weimin, Guan, Hongwen, Ma. 2015. "Polyethylene Glycol/Mesoporous Calcium Silicate Shape-Stabilized Composite Phase Change Material: Preparation, Characterization, and Adjustable Thermal Property." *Energy* 82: 333–40.

[126] Sari, A. 2014. "Composites of Polyethylene Glycol (PEG600) with Gypsum and Natural Clay as New Kinds of Building PCMs for Low Temperature-Thermal Energy Storage." *Energ. Buildings* 69: 184–92.

[127] Sari, A. 2016. "Thermal Energy Storage Characteristics of Bentonite-Based Composite PCMs with Enhanced Thermal Conductivity as Novel Thermal Storage Building Materials." *Energy Convers. Manag.* 117: 132–41.

[128] Onder, E, Sarier, N, Ukuser, G, Ozturk, M, Arat, R. 2013. "Ultrasound Assisted Solvent Free Intercalation of Montmorillonite with PEG1000: A New Type of Organoclay with Improved Thermal Properties." *Thermochim. Acta* 566: 24–35.

[129] Bahramian, AR, Ahmadi, LS, Kokabi, M. 2014. "Performance Evaluation of Polymer/Clay Nanocomposite Thermal Protection Systems Based on Polyethylene Glycol Phase Change Material." *Iranian Polym. J.* 23 (3): 163–69.

[130] Qian, T, Li, Jinhong, Ma, Hongwen, and Yang, Jing. 2016. "Adjustable Thermal Property of Polyethylene Glycol/Diatomite Shape-Stabilized Composite Phase Change Material." *Polymer Composites* 37 (3): 854–60.

[131] Karaman, S, Karaipekli, A, Sarı, A, Biçer, A. 2011. "Polyethylene Glycol (PEG)/Diatomite Composite as a Novel Form-Stable Phase Change Material for Thermal Energy Storage." *Sol. Energy Mater. Sol. Cells* 95 (7): 1647–53.

[132] Qian, T, Li, J, Min, X, Deng, Y, Guan, W, Ning, L. 2015. "Diatomite: A Promising Natural Candidate as Carrier Material for Low, Middle and High Temperature Phase Change Material." *Energy Convers. Manag.* 98: 34–45.

[133] Qian, T, Li, J, Feng, W, Nian, H. 2017. "Enhanced Thermal Conductivity of Form-Stable Phase Change Composite with Single-Walled Carbon Nanotubes for Thermal Energy Storage." *Sci. Rep.* 7 (1): 44710.

[134] Qian, T, Li, J, Min, X, Guan, W, Deng, Y, Ning, L. 2015. "Enhanced Thermal Conductivity of PEG/Diatomite Shape-

Stabilized Phase Change Materials with Ag Nanoparticles for Thermal Energy Storage." *J. Mater. Chem. A* 3 (16): 8526–36.

[135] Qian, Y, Wei, P, Jiang, P, Li, Z, Yan, Y, Liu, J. 2013. "Preparation of a Novel PEG Composite with Halogen-Free Flame Retardant Supporting Matrix for Thermal Energy Storage Application." *Appl. Energy* 106: 321–327.

[136] Gutierrez, A, Ushak, S, Galleguillos, H, Fernandez, A, Cabeza, LF, Grágeda, M. 2015. "Use of Polyethylene Glycol for the Improvement of the Cycling Stability of Bischofite as Thermal Energy Storage Material." *Appl. Energy* 154: 616–21.

[137] Feng, L, Song, P, Yan, Wang, H, Wang, J. 2015. "The Shape-Stabilized Phase Change Materials Composed of Polyethylene Glycol and Graphitic Carbon Nitride Matrices." *Thermochim. Acta* 612: 19–24.

[138] Deng, Y, Li, J, Qian, T, Guan, W, Li, Y, Yin, X. 2016. "Thermal Conductivity Enhancement of Polyethylene Glycol/Expanded Vermiculite Shape-Stabilized Composite Phase Change Materials with Silver Nanowire for Thermal Energy Storage." *Chem. Eng. J.* 295: 427–35..

[139] Deng, Y, Li, J, Nian, H. 2018. "Polyethylene Glycol-Enwrapped Silicon Carbide Nanowires Network/Expanded Vermiculite Composite Phase Change Materials: Form-Stabilization, Thermal Energy Storage Behavior and Thermal Conductivity Enhancement." *Sol. Energy Mater. Sol. Cells* 174: 283–91.

[140] Deng, Y, Li, J, Nian, H, Li, Y, Yin, X. 2017. "Design and Preparation of Shape-Stabilized Composite Phase Change Material with High Thermal Reliability via Encapsulating Polyethylene Glycol into Flower-like TiO_2 nanostructure for Thermal Energy Storage." *Appl. Therm. Eng.* 114: 328–36.

[141] Yang, J, Tang, LS, Bao, RY, Bai, L, Liu, ZY, Xie, BH, Yang, MB, Yang, W. 2018. "Hybrid Network Structure of Boron Nitride and Graphene Oxide in Shape-Stabilized Composite Phase Change Materials with Enhanced Thermal Conductivity and Light-to-

Electric Energy Conversion Capability." *Sol. Energy Mater. Sol. Cells* 174: 56–64..

[142] Su, JC, Liu, PS. 2006. "A Novel Solid–solid Phase Change Heat Storage Material with Polyurethane Block Copolymer Structure." *Energy Convers. Manag.* 47 (18–19): 3185–91.

[143] Li, WD, Ding, EY. 2007. "Preparation and Characterization of Cross-Linking PEG/MDI/PE Copolymer as Solid–solid Phase Change Heat Storage Material." *Sol. Energy Mater. Sol. Cells* 91 (9): 764–68.

[144] Zhou, XM. 2009. "Preparation and Characterization of PEG/MDI/PVA Copolymer as Solid-Solid Phase Change Heat Storage Material." *J. Appl. Polym. Sci.* 113 (3): 2041–45.

[145] Peng, K, Chen, C, Pan, W, Liu, W, Wang, Z, Zhu, L. 2016. "Preparation and Properties of β-Cyclodextrin/ 4,4′-Diphenylmethane Diisocyanate/Polyethylene Glycol (β-CD/MDI/PEG) Crosslinking Copolymers as Polymeric Solid–solid Phase Change Materials." *Sol. Energy Mater. Sol. Cells* 145: 238–47.

[146] Fu, X, Kong, W, Zhang, Y, Jiang, L, Wang, J, Lei, J. 2015. "Novel Solid-Solid Phase Change Materials with Biodegradable Trihydroxy Surfactant for Thermal Energy Storage." *RSC Adv.* 5: 68881–89.

[147] Chen, C, Liu, W, Wang, H, Peng, K. 2015. "Synthesis and Performances of Novel Solid–solid Phase Change Materials with Hexahydroxy Compounds for Thermal Energy Storage." *Appl. Energy* 152: 198–206.

[148] Xi, P, Duan, Y, Fei, P, Xia, L, Liu, R, Cheng, B. 2012. "Synthesis and Thermal Energy Storage Properties of the Polyurethane Solid–solid Phase Change Materials with a Novel Tetrahydroxy Compound." *Eur. Polym. J.* 48 (7): 1295–303.

[149] Xi, P, Zhao, F, Fu, P, Wang, X, Cheng, B. 2014. "Synthesis, Characterization, and Thermal Energy Storage Properties of a Novel Thermoplastic Polyurethane Phase Change Material." *Mater. Lett.* 121: 15–18.

[150] Alkan, C, Günther, E, Hiebler, S, Ensari, OF, Kahraman, D. 2012. "Polyurethanes as Solid–solid Phase Change Materials for Thermal Energy Storage." *Sol. Energy* 86 (6): 1761–69.

[151] Liu, Z, Fu, X, Jiang, L, Wu, B, Wang, J, Lei, J. 2016. "Solvent-Free Synthesis and Properties of Novel Solid–solid Phase Change Materials with Biodegradable Castor Oil for Thermal Energy Storage." *Sol. Energy Mater. Sol. Cells* 147: 177–84.

[152] Du, X, Wang, H, Cheng, X, Du, Z. 2016. "Synthesis and Thermal Energy Storage Properties of a Solid–solid Phase Change Material with a Novel Comb-Polyurethane Block Copolymer Structure." *RSC Adv.* 6 (48): 42643–48. .

[153] Fu, X, Xiao, Y, Hu, K, Wang, J, Lei, J, Zhou, C. 2016. "Thermosetting Solid–solid Phase Change Materials Composed of Poly(Ethylene Glycol)-Based Two Components: Flexible Application for Thermal Energy Storage." *Chem. Eng. J.* 291: 138–48.

[154] Chen, K, Liu, R, Zou, C, Shao, Q, Lan, Y, Cai, X, Zhai, L. 2014. "Linear Polyurethane Ionomers as Solid–solid Phase Change Materials for Thermal Energy Storage. Sol. Energy Mater." *Sol. Cells* 130: 466–73.

[155] Shi, H, Li, J, Jin, Y, Yin, Y, Zhang, X. 2011. "Preparation and Properties of Poly(Vinyl Alcohol)-g-Octadecanol Copolymers Based Solid-Solid Phase Change Materials." *Mater. Chem. Phys.* 131 (1–2): 108–12.

[156] Zhou, Y, Sheng, D, Liu, X, Lin, C, Ji, F, Dong, L, Xu, S, Yang, Y. 2018. "Synthesis and Properties of Crosslinking Halloysite Nanotubes/Polyurethane-Based Solid-Solid Phase Change Materials." *Sol. Energy Mater. Sol. Cells* 174: 84–93.

[157] Cao, Q, Liu, P. 2006. "Hyperbranched Polyurethane as Novel Solid–solid Phase Change Material for Thermal Energy Storage." *Eur. Polym. J.* 42 (11): 2931–39.

[158] Cao, Q, Liu, P. 2007. 'Crystalline-Amorphous Phase Transition of Hyperbranched Polyurethane Phase Change Materials for Energy Storage." *J. Mater. Sci.* 42 (14): 5661–65.

[159] Cao, Q, Liao, L, Xu, H. 2010. "Study on the Influence of Thermal Characteristics of Hyperbranched Polyurethane Phase Change Materials for Energy Storage." *J. Appl. Polym. Sci.* 115 (4): 2228–35.

[160] Liao, L, Cao, Q, Liao, H. 2010. "Investigation of a Hyperbranched Polyurethane as a Solid-State Phase Change Material." *J. Mater. Sci.* 45 (9): 2436–41.

[161] Du, X, Wang, H, Wu, Y, Du, Z, Cheng, X. 2017. "Solid-Solid Phase-Change Materials Based on Hyperbranched Polyurethane for Thermal Energy Storage." *J. Appl. Polym. Sci.* 134 (26): 1–8.

[162] Sundararajan, S, Samui, AB, Kulkarni, PS. 2017. "Thermal Energy Storage Using Poly(Ethylene Glycol) Incorporated Hyperbranched Polyurethane as Solid–Solid Phase Change Material." *Ind. Eng. Chem. Res.* 56 (49): 14401–09.

[163] Li, Y, Liu, R, Huang, Y. 2008. "Synthesis and Phase Transition of Cellulose- Graft -Poly(Ethylene Glycol) Copolymers." *J. Appl. Polym. Sci.* 110 (3): 1797–803.

[164] Li, Y, Wu, M, Liu, R, Huang, Y. 2009. "Cellulose-Based Solid–solid Phase Change Materials Synthesized in Ionic Liquid. Sol Energy Mater." *Sol. Cells* 93 (8): 1321–28.

[165] Kumar, A, Kulkarni, PS, Samui, AB. 2014. "Polyethylene Glycol Grafted Cotton as Phase Change Polymer." *Cellulose* 21 (1): 685–96.

[166] Xi, P, Gu, X, Cheng, B, Wang, Y. 2009. "Preparation and Characterization of a Novel Polymeric Based Solid–solid Phase Change Heat Storage Material." *Energy Convers. Manag.* 50 (6): 1522–28.

[167] Guo, J, Xiang, H, Wang, Q, Hu, C, Zhu, M, Li, L. 2012. "Preparation of Poly(Decaglycerol-Co-Ethylene Glycol) Copolymer as Phase Change Material." *Energ. Buildings* 48: 206–10.

[168] Tang, B, Yang, Z, Zhang, S. 2012. "Poly(Polyethylene Glycol Methyl Ether Methacrylate) as Novel Solid-Solid Phase Change Material for Thermal Energy Storage." *J. Appl. Polym. Sci.* 125 (2): 1377–81.

[169] Zhang, H, Sun, D, Wang, Q, Guo, J, Gong, Y. 2014. "Synthesis and Characterization of Polyethylene Glycol Acrylate Crosslinking Copolymer as Solid-Solid Phase Change Materials." *J. Appl. Polym. Sci.* 131 (6): 39755.

[170] Sundararajan, S, Samui, AB, Kulkarni, PS. 2018. "Synthesis and Characterization of Poly(Ethylene Glycol) Acrylate (PEGA) Copolymers for Application as Polymeric Phase Change Materials (PCMs)." *React. Funct. Polym.* 130: 43–50.

[171] Meng, JY, Tang, XF, Li, W, Shi, HF, Zhang, XX. 2013. "Crystal Structure and Thermal Property of Polyethylene Glycol Octadecyl Ether." *Thermochim. Acta* 558: 83–6.

[172] Meng, JY, Tang, XF, Zhang, ZL, Zhang, XX, Shi, HF. 2013. "Fabrication and Properties of Poly(Polyethylene Glycol Octadecyl Ether Methacrylate)." *Thermochim. Acta* 574: 116–20.

[173] Hu, J, Yu, H, Chen, Y, Zhu, M. 2006. "Study on Phase-Change Characteristics of PET-PEG Copolymers." *J. Macromol. Sci. Part B* 45 (4): 615–21.

[174] Zhou, XM. 2010. "Study on Phase Change Characteristics of PEG/PAM Coupling Blend." *J. Appl. Polym. Sci.* 21 (7): 1591–95.

[175] Gu, X, Xi, P, Cheng, B, Niu, S. 2010. "Synthesis and Characterization of a Novel Solid-Solid Phase Change Luminescence Material." *Polym. Int.* 59 (6): 772–77.

[176] Guo, J, Xie, P, Zhang, X, Yu, C, Guan, F, Liu, Y. 2014. "Synthesis and Characterization of Graft Copolymer of Polyacrylonitrile- g - Polyethylene Glycol-Maleic Acid Monoester Macromonomer." *J. Appl. Polym. Sci.* 131 (8): 40152.

[177] Mu, S-Y, Guo, J, Gong, Y-M, Zhang, S, Yu, Y. 2015. "Synthesis and Thermal Properties of Poly(Styrene-Co-Acrylonitrile)-Graft-Polyethylene Glycol Copolymers as Novel Solid–solid Phase Change Materials for Thermal Energy Storage." *Chinese Chem. Lett.* 26 (11): 1364–66.

[178] Li, Y, Wang, S, Liu, H, Meng, F, Ma, H, Zheng, W. 2014. "Preparation and Characterization of Melamine/Formaldehyde/

Polyethylene Glycol Crosslinking Copolymers as Solid–solid Phase Change Materials." *Sol. Energy Mater. Sol. Cells* 127: 92–7.

[179] Chen, C, Liu, W, Yang, H, Zhao, Y, Liu, S. 2011. "Synthesis of Solid–solid Phase Change Material for Thermal Energy Storage by Crosslinking of Polyethylene Glycol with Poly (Glycidyl Methacrylate)." *Sol. Energy* 85 (11): 2679–85.

[180] Sari, A, Alkan, C, Biçer, A. 2012. "Synthesis and Thermal Properties of Polystyrene-Graft-PEG Copolymers as New Kinds of Solid–solid Phase Change Materials for Thermal Energy Storage." *Mater. Chem. Phys.* 133 (1): 87–94.

[181] Sari, A, Alkan, C, Biçer, A, Karaipekli, A. 2011. "Synthesis and Thermal Energy Storage Characteristics of Polystyrene-Graft-Palmitic Acid Copolymers as Solid–solid Phase Change Materials." *Sol. Energy Mater. Sol. Cells* 95 (12): 3195–201.

[182] Xiang, H, Wang, S, Wang, R, Zhou, Z, Peng, C, Zhu, M. 2013. "Synthesis and Characterization of an Environmentally Friendly PHBV/PEG Copolymer Network as a Phase Change Material." *Sci. China Chem.* 56 (6): 716–23.

[183] Sundararajan, S, Samui, AB, Kulkarni, PS. 2016. "Interpenetrating Phase Change Polymer Networks Based on Crosslinked Polyethylene Glycol and Poly(Hydroxyethyl Methacrylate)." *Sol. Energy Mater. Sol. Cells* 149: 266–74.

[184] Vigo, TL, Frost, CM. 1985. "Temperature-Adaptable Fabrics." *Text. Res. J.* 55 (12): 737–43.

[185] Karthikeyan, M, Ramachandran, T, Sundaram, OLS. 2013. "Nanoencapsulated Phase Change Materials Based on Polyethylene Glycol for Creating Thermoregulating Cotton." *J. Ind. Text.* 44 (1): 130–46.

[186] Kuru, A, Aksoy, SA. 2014. "Cellulose-PEG Grafts from Cotton Waste in Thermo-Regulating Textiles." *Text. Res. J.* 84 (4): 337–46.

[187] Khoddami, A, Avinc, O, Ghahremanzadeh, F. 2011. "Improvement in Poly(Lactic Acid) Fabric Performance via Hydrophilic Coating." *Prog. Org. Coat.* 72 (3): 299–304.

[188] Ghahremanzadeh, F, Khoddami, A, Carr, CM. 2010. "Improvement in Fastness Properties of Phase-Change Material Applied on Surface Modified Wool Fabrics." *Fibers and Polymers* 11 (8): 1170–80.

[189] Nguyen, TTT, Lee, JG, Park, JS. 2011. "Fabrication and Characterization of Coaxial Electrospun Polyethylene Glycol/Polyvinylidene Fluoride (Core/Sheath) Composite Non-Woven Mats." *Macromol. Res.* 19 (4): 370–78.

[190] Nguyen, TTT, Park, JS. 2011. "Fabrication of Electrospun Nonwoven Mats of Polyvinylidene Fluoride/Polyethylene Glycol/Fumed Silica for Use as Energy Storage Materials." *J. Appl. Polym. Sci.* 121 (6): 3596–603.

[191] Ke, GZ, Xie, HF, Ruan, RP, Yu, WD. 2010. "Preparation and Performance of Porous Phase Change Polyethylene Glycol/Polyurethane Membrane." *Energy Convers. Manag.* 51 (11): 2294–98.

[192] Baetens, RR, Jelle, BP, Gustavsen, A. 2010. "Phase Change Materials for Building Applications: A State-of-the-Art Review." *Energy and Build.* 42 (9): 1361–68.

[193] Tang, B, Wang, Y, Qiu, M, Zhang, S. 2014. "A Full-Band Sunlight-Driven Carbon Nanotube/PEG/SiO2 Composites for Solar Energy Storage." *Sol. Energy Mater. Sol. Cells* 123: 7–12.

[194] Wang, Y, Bingtao, T, Shufen, Z. 2013. "Single-Walled Carbon Nanotube/Phase Change Material Composites: Sunlight-Driven, Reversible, Form-Stable Phase Transitions for Solar Thermal Energy Storage." *Adv. Funct. Mater.* 23 (35): 4354–60.

[195] Chen, C, Liu, W, Wang, Z, Peng, K, Pan, W, Xie, Q. 2015. "Novel Form Stable Phase Change Materials Based on the Composites of Polyethylene Glycol/Polymeric Solid-Solid Phase Change Material." *Sol. Energy Mater. Sol. Cells* 134: 80–8.

[196] Skach, M, Arora, M, Hsu, CH, Li, Q, Tullsen, D, Tang, L, Mars, J. 2015. "Thermal Time Shifting." In *Proceedings of the 42nd Annual International Symposium on Computer Architecture - ISCA '15*, 439–49. New York, New York, USA: ACM Press.

[197] Deign, J. 2016. *How Phase-Change Materials Are Changing Lives.* http://energystoragereport.info/phase-change-materials-saving-lives/.

[198] Formato, RM. 2013. "The Advantages & Challenges of Phase Change Materials (PCMs)." In *Thermal Packaging Author.* http://www.coldchaintech.com/assets/Cold-Chain-Technologies-PCM-White-Paper.pdf.

[199] Andrews, JR, Prajapati, KG, Eypper, E, Shrestha, P, Shakya, M, Pathak, KR, Joshi, N, Tiwari, P, Risal, M, Koirala, S, Karkey, A, Dongol, S, Wen, S, Smith, AB, Maru, D, Basnyat, B, Baker, S, Farrar, J, Ryan, ET, Hohmann, E, Arjyal, A. 2013. "Evaluation of an Electricity-Free, Culture-Based Approach for Detecting Typhoidal Salmonella Bacteremia during Enteric Fever in a High Burden, Resource-Limited Setting." *PLoS: Neglected Tropical Diseases* 7 (6): e2292.

[200] Kim, H, Lee, D, Kim, J, Kim, T, Kim, WJ. 2013. "Photothermally Triggered Cytosolic Drug Delivery via Endosome Disruption Using a Functionalized Reduced Graphene Oxide." *ACS Nano* 7 (8): 6735–46.

[201] Duong, B, Liu, H, Ma, L, Su, M. 2014. "Covert Thermal Barcodes Based on Phase Change Nanoparticles." *Sci. Rep.* 4: 5170.

[202] Bhamidipati, MV, 2008 Nov 13. "Methods and compositions for inhibiting surface icing." *United States Patent* US 20100322867A1.

[203] Maxa, J, Novikov, A, Nowottnick, M. 2018. "Thermal Peak Management Using Organic Phase Change Materials for Latent Heat Storage in Electronic Applications." *Materials* 11 (1): 31.

[204] Rigotti, D, Dorigato, A, Pegoretti, A. 2018. "3D Printable Thermoplastic Polyurethane Blends with Thermal Energy Storage/Release Capabilities." *Mater. Today Commun.* 15: 228–35.

[205] Huang, Y-L, Li, Q-B, Deng, X, Lu, Y-H, Liao, X-K, Hong, M-Y, Wang, Y. 2005. "Aerobic and Anaerobic Biodegradation of Polyethylene Glycols Using Sludge Microbes." *Process Biochem.* 40 (1): 207–11.

In: Phase Change Materials
Editor: Ismaël van der Winden

ISBN: 978-1-53617-536-3
© 2020 Nova Science Publishers, Inc.

Chapter 2

COMPACT RECONFIGURABLE OPTICAL DEVICES USING PHASE-CHANGE MATERIALS

Yin Huang[1,*], *Lanyan Wang*[1], *Yuecheng Shen*[2], *Changjun Min*[3] *and Georgios Veronis*[4,5]

[1]Department of Optoelectrics Information Science and Engineering,
School of Physics and Electronics, Central South University,
Changsha, Hunan, China
[2]Department of Medical Engineering,
California Instituteof Technology, Pasadena, CA, US
[3]Key Laboratory of Optoelectronic Devices
and Systems of Ministry of Education and Guangdong Province,
Shenzhen University, Shenzhen, China
[4]School of Electrical Engineering and Computer Science,
Louisiana State University, BatonRouge, LA, US
[5]Center for Computation and Technology,
Louisiana State University, BatonRouge, LA, US

Abstract

We introduce a non-parity-time-symmetric three-layer structure, consisting of a gain medium layer sandwiched between two phase-change medium layers for switching of the direction of reflectionless light propagation. We show that for this structure unidirectional reflectionlessness in the forward direction can be switched to unidirectional reflectionlessness in the backward direction at the optical communication wavelength by switching the phase-change material $Ge_2Sb_2Te_5$ (GST) from its amorphous to its crystalline phase. We also show that it is the existence of exceptional points for this structure with GST in both its amorphous and crystalline phases which leads to unidirectional reflectionless propagation in the forward direction for GST in its amorphous phase, and in the backward direction for GST in its crystalline phase. We further switch photonic nanostructures between cloaking and superscattering regimes using phase-change materials at mid-infrared wavelengths. More specifically, we investigate the scattering properties of subwavelength three-layer cylindrical structures in which the material in the outer shell is the phase-change material GST. We first show that, when GST is switched between its amorphous and crystalline phases, properly designed electrically small structures can switch between resonant scattering and cloaking invisibility regimes. The contrast ratio between the scattering cross sections of the cloaking invisibility and resonant scattering regimes reaches almost unity. We then also show that larger, moderately small cylindrical structures can be designed to switch between superscattering and cloaking invisibility regimes, when GST is switched between its crystalline and amorphous phases. The contrast ratio between the scattering cross sections of cloaking invisibility and superscattering regimes can be as high as ~93%.

1. Introduction

In recent years, investigating the interaction of light with subwavelength structures has attracted a lot of attention, since it could potentially lead to a new generation of photonic devices [1–6]. In particular, the capability to control the scattering of light and achieve invisibility cloaking of subwavelength structures is important for applications in biomedicine, photovoltaics, sensing, optical detection, and near-field imaging [7–13]. In the past few years, the use of plasmonic and dielectric multilayer coatings to drastically reduce the total scattering cross-section of deep subwavelength objects, and thus achieve invisibility cloaking based on scattering cancellation, has been explored [14–18]. In addition to cloaking, it has been demonstrated that subwavelength multilayer core-shell structures can lead to enhanced resonant scattering, so that the scattering cross sections of the original structures are greatly enhanced [19–22]. This phenomenon is commonly referred to as superscattering. Switching between the cloaking and enhanced scattering states could be essential for building compact optoelectronic devices, for reducing the size of optical systems, and for developing reconfigurable optical components [23–25]. Such switching between the enhanced scattering and invisibility cloaking regimes has been demonstrated using nonlinear materials [26] and quantum emitters [27]. An alternative way to achieve this switching could be through the use of materials with tunable optical properties, such as phase-change materials.

Ge2Sb2Te5 (GST) is a phase-change material with an amorphous and a crystalline phase [28]. The covalently bonded amorphous phase of GST corresponds to a disordered material with short-range atomic order. In contrast, the resonantly bonded crystalline phase can be regarded as a semiconductor with orderly aligned atoms. Thus, the optical properties of amorphous GST (aGST) and crystalline GST (cGST) are significantly different. These two phases can be switched reversibly and rapidly by applying external electrical pulses, laser pulses or thermal annealing. Picosecond-order crystallization times have been reported for GST by femtosecond laser pulses [29, 30]. Amorphization of GeSbTe has been achieved on subpicosecond timescales with femtosecond laser pulse excitation [31]. In addition to being inexpensive and easy to use in device fabrication processes, GST retains its phase for years after removal of the external excitations. GST has been widely used for non-volatile, rewritable optical data storage devices and for electronic memories [32, 33]. Recently, it has been shown that GST could provide a versatile platform for the realization of optically reconfigurable active photonic devices due to its switchable dielectric properties [34–40].

We investigate the scattering properties of a three-layer cylindrical structure with GST at the mid-infrared wavelength of 4 µm. We first consider an electrically small structure. We show that, when GST is switched between its amorphous and crystalline phases, the structure switches between resonant scattering and cloaking invisibility regimes. The contrast ratio between the scattering cross sections of the cloaking invisibility and resonant scattering regimes reaches almost unity. We then consider the case of a larger, moderately small cylindrical structure. In this scenario, we demonstrate that, when GST is switched between its crystalline and amorphous phases, the structure switches between superscattering and cloaking invisibility regimes. The contrast ratio between the scattering cross sections of cloaking invisibility and superscatteringregimes can be as high as ~93%. Although here we focus on two-dimensional infinitely longcylindrical structures, the proposed approach is rather general and can be applied to other optical structures.

Exceptional points, which are branch point singularities of the spectrum, are associated with the coalescence of both eigenvalues and corresponding eigenstates in open quantum systems described by non-Hermitian Hamiltonians [41-47]. Exceptional points have been studied in lasers [48], coupled dissipative dynamical systems [49], mechanics [50], electronic circuits [51], and atomic as well as molecular systems [52]. In the past few years, unidirectional light reflectionlessness caused by the existence of exceptional points in non-Hermitian parity-time (PT) symmetric optical systems possessing balanced gain and loss has attracted considerable attention [53-57]. In such structures the reflection is zero when measured from one end of the structure at optical exceptional points, and nonzero when measured from the other end. Unidirectional light reflectionlessness can also be attained in non-PT-symmetric structures with unbalanced gain and loss [6, 58-69]. This is due to the fact that exceptional points exist in a larger family of non-Hermitian Hamiltonians [58]. Achieving unidirectional reflectionless propagation is important for several key applications in photonic circuits such as optical network analyzers [55, 59]. In addition, switching of the direction of reflectionless light propagation could be essential for building compact optoelectronic devices, for reducing the size of optical systems, and for developing reconfigurable optical components [23–25]. This could be achieved by using materials with tunable optical properties such as phase-change materials.

Motivated by the transport behavior enabled by non-Hermiticity and the high refractive index contrast between the amorphous and crystalline phases of phase-change material GST, we use a non-PT-symmetric three-layer structure, consisting of a gain medium layer sandwiched between two GST layers, to switch the direction of reflectionless light propagation at exceptional points. We show that, when GST is switched from its amorphous to its crystalline phase, the structure switches from unidirectional reflectionless in the forward direction tounidirectional reflectionless in the backward direction. The structure is designed at the optical communication wavelength of $\lambda_0 = 1.55\mu$m. We then discuss the underlying physical mechanism of unidirectional reflectionless light propagation in this structure. We show that a layer with gain has to be included in the structure to compensate the loss in the GST layers so as to achieve complete destructive interference. We demonstrate that the structure exhibits exceptional points for GST in both its amorphous and crystalline phases. These exceptional points result inunidirectional reflectionless propagation in the forward direction for GST in its amorphous phase, and in the backward direction for GST in its crystalline phase. We investigate the phase transitions associated with the exceptional points. Finally, the topological structure of the exceptional points is also explored by encircling them in parameter space.

2. SWITHING PHOTONIC NANOSTRUCTURES BETWEEN CLOAKING AND SUPERCATTERING

2.1. Theory

Our proposed three-layer cylindrical structure is normally illuminated by a TM plane wave propagating in the x direction with the magnetic field polarized along the cylinder (z) axis, as illustrated in Figure 1. The material in the outer shell is the phase-change material GST. Based on the Mie-Lorenz mode-expansion method [73–75], the expansions for the incident fields are given by

$$H_{in} = \sum_{n=-\infty}^{\infty} H_n N_n^{(1)} \tag{1}$$

$$E_{in} = \frac{ik_0}{\omega \varepsilon_0} \sum_{n=-\infty}^{\infty} H_n M_n^{(1)} \tag{2}$$

with expansion coefficients

$$H_n = \frac{H_0(-i)^n}{k_0} \tag{3}$$

where H_0 is the strength of the incident magnetic field, k0 is the wave number in free space, and E_0 is the dielectric permittivity of free space. M_n and N_n are vector cylindrical harmonics of the n-th order [73]. The superscript (1) indicates that for the vector cylindrical harmonics the radial dependence of the fields is given by Bessel functions of the first kind J_n. The expansions of the fields in the core layer (Figure 1) are

$$H_1 = \sum_{n=-\infty}^{\infty} H_n \left[ic_n M_n^{(1)} + d_n N_n^{(1)} \right] \tag{4}$$

$$E_1 = \frac{ik_1}{\omega \epsilon_1} \sum_{n=-\infty}^{\infty} H_n \left[ic_n N_n^{(1)} + d_n M_n^{(1)} \right] \tag{5}$$

where k_1 is the wave number in the core layer, and ϵ_1 is the dielectric permittivity of the material in the core. The expansions of the fields in the inner shell with dielectric permittivity ϵ_2 (Figure 1) can be expressed as

$$H_2 = \sum_{n=-\infty}^{\infty} H_n \left[ig_n M_n^{(1)} + f_n N_n^{(1)} + ip_n M_n^{(2)} + q_n N_n^{(2)} \right] \tag{6}$$

$$E_2 = \frac{ik_2}{\omega \epsilon_2} \sum_{n=-\infty}^{\infty} H_n \left[ig_n N_n^{(1)} + f_n M_n^{(1)} + ip_n N_n^{(2)} + q_n M_n^{(2)} \right] \tag{7}$$

where k_2 is the wave number in the inner shell region. The superscript (2) indicates that for the vector cylindrical harmonics the radial dependence of the fields is given by Bessel functions of the second kind Y_n. Similarly, the expansions of the fields in the outer shell region can be written as

$$H_3 = \sum_{n=-\infty}^{\infty} H_n \left[is_n M_n^{(1)} + t_n N_n^{(1)} + iw_n M_n^{(2)} + v_n N_n^{(2)} \right] \quad (8)$$

$$E_3 = \frac{ik_3}{\omega \epsilon_3} \sum_{n=-\infty}^{\infty} H_n \left[is_n N_n^{(1)} + t_n M_n^{(1)} + iw_n N_n^{(2)} + v_n M_n^{(2)} \right] \quad (9)$$

where k_3 is the wave number in the outer shell, and ϵ_3 is the dielectric constant of GST. The scattered fields outside the three-layer cylindrical structure are given by

$$H_s = \sum_{n=-\infty}^{\infty} H_n \left[ib_n M_n^{(3)} + a_n N_n^{(3)} \right] \quad (10)$$

$$E_s = \frac{ik_0}{\omega \epsilon_0} \sum_{n=-\infty}^{\infty} H_n \left[ib_n N_n^{(3)} + a_n M_n^{(3)} \right] \quad (11)$$

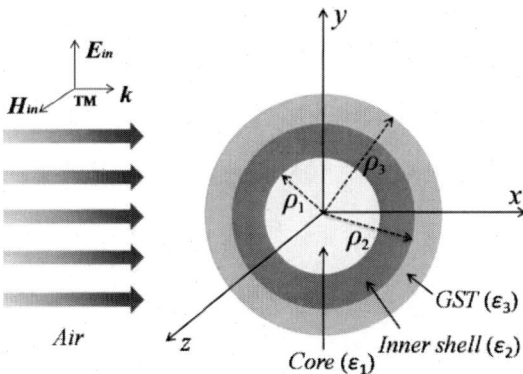

Figure 1. Schematic of a three-layer core-shell cylindrical structure. The material in the outer shell is the phase-change material GST.

The superscript (3) in the above equations indicates that for the vector cylindrical harmonics the radial dependence of the fields is given by Hankel functions of the first kind, H_n. By applying the boundary conditions at $\rho = \rho_j$, $j = 1, 2, 3$, we obtain the scattering coefficients a_n and b_n.

$$a_n = 0 \tag{12}$$

$$b_n = \frac{U_n^{TM}}{U_n^{TM} + iV_n^{TM}} \tag{13}$$

Note that the scattering coefficients an vanish when the plane wave is normally incident on the cylindrical structure [73]. In Eq. (13), U_n^{TM} and V_n^{TM} are given by

$$U_n^{TM} = \begin{vmatrix} J_n(k_1\rho_1) & J_n(k_2\rho_1) & Y_n(k_2\rho_1) & 0 & 0 & 0 \\ \frac{J_n'(k_1\rho_1)}{\eta_1} & \frac{J_n'(k_2\rho_1)}{\eta_2} & \frac{Y_n'(k_2\rho_1)}{\eta_2} & 0 & 0 & 0 \\ 0 & J_n(k_2\rho_2) & Y_n(k_2\rho_2) & J_n(k_3\rho_2) & Y_n(k_3\rho_2) & 0 \\ 0 & \frac{J_n'(k_2\rho_2)}{\eta_2} & \frac{Y_n'(k_2\rho_2)}{\eta_2} & \frac{J_n'(k_3\rho_2)}{\eta_3} & \frac{Y_n'(k_3\rho_2)}{\eta_3} & 0 \\ 0 & 0 & 0 & J_n(k_3\rho_2) & Y_n(k_3\rho_3) & J_n(k_0\rho_3) \\ 0 & 0 & 0 & \frac{J_n'(k_3\rho_2)}{\eta_3} & \frac{Y_n'(k_3\rho_3)}{\eta_3} & \frac{J_n'(k_0\rho_3)}{\eta_0} \end{vmatrix} \tag{14}$$

and

$$V_n^{TM} = \begin{vmatrix} J_n(k_1\rho_1) & J_n(k_2\rho_1) & Y_n(k_2\rho_1) & 0 & 0 & 0 \\ \frac{J_n'(k_1\rho_1)}{\eta_1} & \frac{J_n'(k_2\rho_1)}{\eta_2} & \frac{Y_n'(k_2\rho_1)}{\eta_2} & 0 & 0 & 0 \\ 0 & J_n(k_2\rho_2) & Y_n(k_2\rho_2) & J_n(k_3\rho_2) & Y_n(k_3\rho_2) & 0 \\ 0 & \frac{J_n'(k_2\rho_2)}{\eta_2} & \frac{Y_n'(k_2\rho_2)}{\eta_2} & \frac{J_n'(k_3\rho_2)}{\eta_3} & \frac{Y_n'(k_3\rho_2)}{\eta_3} & 0 \\ 0 & 0 & 0 & J_n(k_3\rho_2) & Y_n(k_3\rho_3) & Y_n(k_0\rho_3) \\ 0 & 0 & 0 & \frac{J_n'(k_3\rho_2)}{\eta_3} & \frac{Y_n'(k_3\rho_3)}{\eta_3} & \frac{Y_n'(k_0\rho_3)}{\eta_0} \end{vmatrix} \tag{15}$$

where $\eta_j = \sqrt{\frac{\varepsilon_j}{\mu_j}}$, $j = 0, 1, 2, 3$. All materials are non-magnetic, so that $\mu_j = \mu_0, j = 1, 2, 3$. The total scattering cross section (SCS), defined as the ratio of the total scattered power to the intensity of the incident plane wave [12, 15, 73], is given by

$$\sigma = \frac{2\lambda_0}{\pi} \sum_{n=-\infty}^{\infty} |b_n|^2 = \frac{2\lambda_0}{\pi} \sigma_N \qquad (16)$$

where λ_0 is the free-space wavelength, and σ_N is the normalized scattering cross section (NSCS) [76]. Based on Eq. (13), $U_n^{TM} = 0$ leads to scattering suppression of the nth ordermultipole. Cloaking invisibility is achieved when all scattering coefficients b_n simultaneously approach zero. On the contrary, resonant scattering of the nth order multipole occurs when $V_n^{TM} = 0$, which provides an opportunity to dramatically enhance this scattering order. Based on electromagnetic duality, the scattering coefficients b_n for a normally incident TE excitation are zero, while $a_n = \frac{U_n^{TE}}{U_n^{TE} + iV_n^{TE}}$, where U_n^{TE} and V_n^{TE} can be readily obtained by replacing ϵ with μ in Eqs. (14) and (15), respectively [15, 73].

2.2. Results

In this section, we use the phase-change material GST in the three-layer cylindrical structure of Figure 1, to switch this structure between cloaking and enhanced scattering regimes at the mid-infrared wavelength of $\lambda_0 = 4 \, \mu m$. Mid-infrared radiation in the μm wavelength range can propagate through several materials without significant intensity attenuation. Because of this property, there is a wide range of potential military and civil applications in the mid-infrared wavelength regime [77]. For the material in the core of the three-layer cylindrical structure (Figure 1) we choose zinc oxide (ZnO). Zinc oxide microwires and nanowires have great potential to be used in many commercial applications due to their low cost and simple fabrication process [78–80]. We use the commercial software COMSOL, which is based on the finite-element method, to numerically calculate the SCS of the proposed structures.

2.2.1. Electrically Small Cylindrical Structures

We first consider electrically small cylindrical structures ($k_0\rho_3 \ll 1$). In this case, by using the asymptotic forms of the Bessel functions, the expressions in Eqs. (14) and (15) for $n \neq 0$ can be reduced to the following [81]

$$U_n^{TM} \simeq \frac{\pi(k_0\rho_3)^n}{4^n n!(n-1)!} \begin{vmatrix} 1 & 1 & -1 & 0 & 0 & 0 \\ \frac{1}{\epsilon_1} & \frac{1}{\epsilon_2} & \frac{1}{\epsilon_2} & 0 & 0 & 0 \\ 0 & \left(\frac{\rho_2}{\rho_1}\right)^n & -\left(\frac{\rho_1}{\rho_2}\right)^n & 1 & -1 & 0 \\ 0 & \frac{1}{\epsilon_2}\left(\frac{\rho_2}{\rho_1}\right)^n & \frac{1}{\epsilon_2}\left(\frac{\rho_1}{\rho_2}\right)^n & \frac{1}{\epsilon_3} & \frac{1}{\epsilon_3} & 0 \\ 0 & 0 & 0 & \left(\frac{\rho_3}{\rho_2}\right)^n & -\left(\frac{\rho_2}{\rho_3}\right)^n & 1 \\ 0 & 0 & 0 & \frac{1}{\epsilon_3}\left(\frac{\rho_3}{\rho_2}\right)^n & \frac{1}{\epsilon_3}\left(\frac{\rho_2}{\rho_3}\right)^n & \frac{1}{\epsilon_0} \end{vmatrix}$$

(17)

and

$$V_n^{TM} \simeq (k_0\rho_3)^{-n} \begin{vmatrix} 1 & 1 & -1 & 0 & 0 & 0 \\ \frac{1}{\epsilon_1} & \frac{1}{\epsilon_2} & \frac{1}{\epsilon_2} & 0 & 0 & 0 \\ 0 & \left(\frac{\rho_2}{\rho_1}\right)^n & -\left(\frac{\rho_1}{\rho_2}\right)^n & 1 & -1 & 0 \\ 0 & \frac{1}{\epsilon_2}\left(\frac{\rho_2}{\rho_1}\right)^n & \frac{1}{\epsilon_2}\left(\frac{\rho_1}{\rho_2}\right)^n & \frac{1}{\epsilon_3} & \frac{1}{\epsilon_3} & 0 \\ 0 & 0 & 0 & \left(\frac{\rho_3}{\rho_2}\right)^n & -\left(\frac{\rho_2}{\rho_3}\right)^n & -1 \\ 0 & 0 & 0 & \frac{1}{\epsilon_3}\left(\frac{\rho_3}{\rho_2}\right)^n & \frac{1}{\epsilon_3}\left(\frac{\rho_2}{\rho_3}\right)^n & \frac{1}{\epsilon_0} \end{vmatrix}$$

(18)

In addition, if the material in the core is dielectric and the structure is electrically small, the scattering cross section for TM excitation is dominated by the terms involving $b = \pm 1$, which are associated with dipolar scattering, and all other terms with $n \neq 1$ are negligible [15, 82]. Note that $|b_n| = |b_{-n}|$ [73]. Thus, in this case, by setting $U_1^{TM} = 0$ in Eq. (17), we obtain the following condition to achieve invisibility cloaking

$$\phi_u = \gamma_2^2 = \frac{(\epsilon_3+\epsilon_0)[(\epsilon_2-\epsilon_1)(\epsilon_3+\epsilon_2)-(\epsilon_1+\epsilon_2)(\epsilon_2-\epsilon_3)\gamma_1^2]}{(\epsilon_0-\epsilon_3)[(\epsilon_1-\epsilon_2)(\epsilon_3-\epsilon_2)-(\epsilon_1+\epsilon_2)(\epsilon_2+\epsilon_3)\gamma_1^2]} \quad (19)$$

where $\gamma_1 = \frac{\rho_2}{\rho_1}$ and $\gamma_2 = \frac{\rho_3}{\rho_2}$. Similarly, by setting $V_1^{TM} = 0$ in Eq. (18), we obtain the following condition to achieve enhanced scattering

$$\phi_v = \gamma_2^2 = \frac{(\epsilon_0-\epsilon_3)[(\epsilon_2-\epsilon_1)(\epsilon_3+\epsilon_2)-(\epsilon_1+\epsilon_2)(\epsilon_2-\epsilon_3)\gamma_1^2]}{(\epsilon_3+\epsilon_0)[(\epsilon_1-\epsilon_2)(\epsilon_3-\epsilon_2)-(\epsilon_1+\epsilon_2)(\epsilon_2+\epsilon_3)\gamma_1^2]} \quad (20)$$

To realize switching of the three-layer cylindrical structure of Figure 1 between cloaking and enhanced scattering regimes, requires $U_1^{TM} = 0$ when GST is in its crystalline phase with dielectric constant ϵ_{3c}, and also $V_1^{TM} = 0$ when GST is in its amorphous phase with dielectric constant ϵ_{3a}. In other words, when GST is in its crystalline phase with dielectric constant ϵ_{3c}, Eq. (19) should be satisfied, while, when GST is in its amorphous phase with dielectric constant ϵ_{3a}, Eq. (20) should be satisfied. Since the left hand sides of Eqs. (19) and (20) are equal, these two conditions can be simultaneously satisfied if $\phi_u(\epsilon_3 = \epsilon_{3c}) = \phi_v(\epsilon_3 = \epsilon_{3a})$. To achieve this, we optimize γ_1 and the dielectric constant of the inner shell ϵ_2, to satisfy the following relation at the wavelength of $\lambda_0 = 4\mu m$

$$|\phi_u(\epsilon_3 = \epsilon_{3c}) - \phi_v(\epsilon_3 = \epsilon_{3a})| =$$
$$\left|\frac{(\epsilon_{3c}+\epsilon_0)[(\epsilon_2-\epsilon_1)(\epsilon_{3c}+\epsilon_2)-(\epsilon_1+\epsilon_2)(\epsilon_2-\epsilon_{3c})\gamma_1^2]}{(\epsilon_0-\epsilon_{3c})[(\epsilon_1-\epsilon_2)(\epsilon_{3c}-\epsilon_2)-(\epsilon_1+\epsilon_2)(\epsilon_2+\epsilon_{3c})\gamma_1^2]} - \frac{(\epsilon_0-\epsilon_{3a})[(\epsilon_2-\epsilon_1)(\epsilon_{3a}+\epsilon_2)-(\epsilon_1+\epsilon_2)(\epsilon_2-\epsilon_{3a})\gamma_1^2]}{(\epsilon_{3a}+\epsilon_0)[(\epsilon_1-\epsilon_2)(\epsilon_{3a}-\epsilon_2)-(\epsilon_1+\epsilon_2)(\epsilon_2+\epsilon_{3a})\gamma_1^2]}\right| = 0 \quad (21)$$

We use experimental data for the frequency-dependent dielectric constants of aGST and cGST [28]. The refractive indices of aGST and cGST are $n_{aGST} = 4.05$ and $n_{cGST} = 5.9 + 0.16j$, respectively, at $\lambda_0 = 4\mu m$ [28, 29]. Thus, at $\lambda_0 = 4\mu m$ cGST is lossy, while aGST is lossless. The dielectric constant of zinc oxide, which is the material used in the core (Figure 1), is $\epsilon_1 = 8.15$ at $\lambda_0 = 4\mu m$.

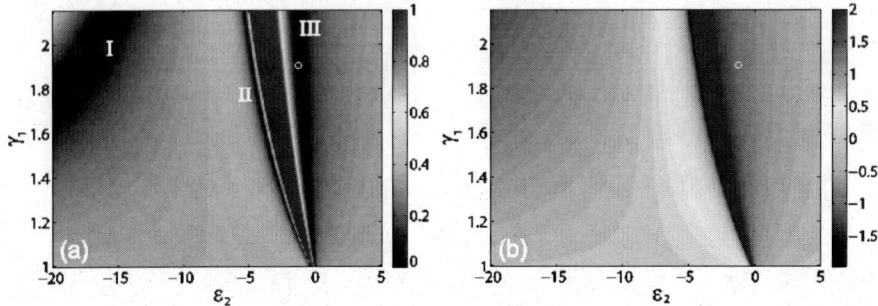

Figure 2. (a) The quantity $|\phi_u(\epsilon_3 = \epsilon_{3c}) - \phi_v(\epsilon_3 = \epsilon_{3a})|$ as a function of $\gamma_1 = \frac{\rho_2}{\rho_1}$ and the dielectric constant of the inner shell ϵ_2 (Figure 1). Results are shown for $\lambda_0 = 4\mu m$. Eq. (21) is satisfied at three regions in the $\gamma_1 - \epsilon_2$ space, marked as I, II, and III. The open circle corresponds to a point in region III. (b) The quantity $\gamma_2^2 = \left(\frac{\rho_3}{\rho_2}\right)^2$ as a function of $\gamma_1 = \frac{\rho_2}{\rho_1}$ and the dielectric constant of the inner shell ϵ_2 (Figure 1). Results are shown for $\lambda_0 = 4\mu m$.

We first neglect the loss of cGST at $\lambda_0 = 4\mu m$. In other words, we assume that $\epsilon_{3a} = 4.05^2 = 16.4$ and $\epsilon_{3c} = 5.9^2 = 34.81$. In Figure 2(a), we show $|\phi_u(\epsilon_3 = \epsilon_{3c}) - \phi_v(\epsilon_3 = \epsilon_{3a})|$ as a function of $\gamma_1 = \frac{\rho_2}{\rho_1}$ and the dielectric constant of the inner shell ϵ_2. In Figure 2(a) we observe three regions in the $\gamma_1 - \epsilon_2$ space (marked as I, II, and III) at which Eq. (21) is satisfied. However, since $\gamma_2^2 = \left(\frac{\rho_3}{\rho_2}\right)^2$ only $\gamma_2^2 \geq 1$ corresponds to physical solutions. Figure 2(b) shows $\gamma_2^2 = \left(\frac{\rho_3}{\rho_2}\right)^2$ as a function of $\gamma_1 = \frac{\rho_2}{\rho_1}$ and the dielectric constant of the inner shell ϵ_2. We observe that region II (Figure 2(a)) does not correspond to physical solutions, since γ_2^2 is negative. On the other hand, regions I and III correspond to physical solutions. We consider a point in region III [indicated by an open circle in Figures 2(a) and 2(b)) with $\gamma_1 = \frac{\rho_2}{\rho_1} = 1.9$, $\epsilon_2 = -1.25$, and $\gamma_2 = \frac{\rho_3}{\rho_2} =$

1.1145. Figure 3(a) shows the NSCS σ_N corresponding to this point as a function of the dielectric constant of the outer shell ϵ_3 calculated with COMSOL. The radius of the ZnO core cylinder ρ_1 is set equal to 23 nm, so that the three-layer structure is electrically small ($k_0\rho_3 = k_0\rho_1\gamma_1\gamma_2 < 0.1$) [83]. We observe a peak ($\sigma_N \simeq 2|b_1|^2 = 1.9766$) and a dip ($\sigma_N \simeq 2|b_1|^2 = 1.01 \times 10^{-12}$) in the scattering cross section for $\epsilon_3 \simeq 16.4$ (aGST) and $\epsilon_3 \simeq 34.81$ (cGST), respectively. Thus, the simulation results confirm that for the optimized electrically small cylindrical structure of Figure 1 switching between cloaking and enhanced scattering regimes can be realized at the mid-infrared wavelength of 4 μm by switching the phase-change material GST between its crystalline and amorphous phases. In Figure 3(a), we also observe a resonance with asymmetric Fano-like shape at $\epsilon_3 = 0$. This is due to the strong interference between the cloaking and resonant scattering states within the same structure [83, 84]. Based on Eqs. (19) and (20), the cloaking and resonant scattering conditions coincide with each other when $\epsilon_3 = 0$, and thus a degenerate cloaking-resonant state is formed.

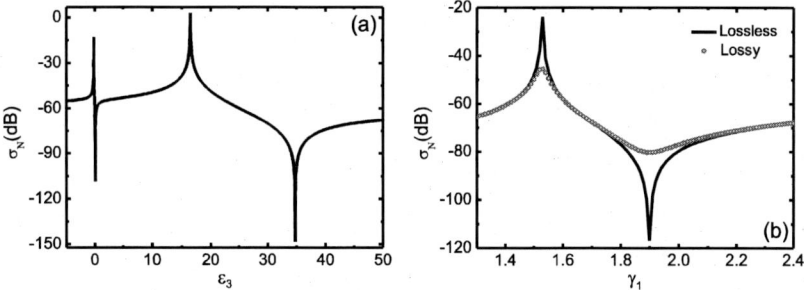

Figure 3. (a) The NSCS σ_N as a function of the dielectric constant of the outer shell ϵ_3 for electrically small cylindrical structures as in Figure 1. Results are shown for $\lambda_0 = 4$ μm, $\rho_1 = 23nm$, $\gamma_1 = \frac{\rho_2}{\rho_1} = 1.9$, $\gamma_2 = \frac{\rho_3}{\rho_2} = 1.1145$ and $\epsilon_2 = -1.25$. The material in the core is ZnO. (b) The NSCS σ_N as a function of when the phase change material GST in theouter shell is in its crystalline phase. The black line and red circles correspond to lossless and lossy cGST, respectively. All other parameters are as in Figure 3(a).

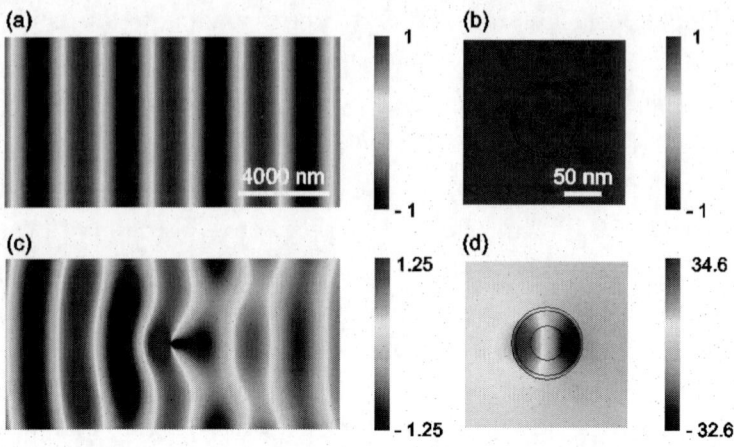

Figure 4 (a) and (b) Magnetic field amplitude profiles for the optimized electrically small structure of Figure 1 with GST in its crystalline phase at $\lambda_0 = 4\ \mu m$, when a plane wave is normally incident from the left. All other parameters are as in Figure 3(a). The fields are normalized with respect to the field amplitude of the incident plane wave. (c) and (d) Same as in (a) and (b) except that GST is in its amorphous phase.

Figure 3(b) shows the NSCS σ_N as a function of $\gamma_1 = \frac{\rho_2}{\rho_1}$ for the optimized structure when the phase-change material GST is in its crystalline phase. Although the material loss in cGST affects the total scattering excursion, cloaking invisibility is still achieved ($\sigma_N = 4.2 \times 10^{-9}$) when $\gamma_1 = \frac{\rho_2}{\rho_1} = 1.9$ (red circles). The contrast ratio between the cloaking and resonant scattering state, define as $\tau \equiv \left|\frac{\sigma_N|aGST - \sigma_N|cGST}{\sigma_N|aGST + \sigma_N|cGST}\right|$, is almost unity. The switching between cloaking and enhancedscattering regimes can be observed in the field distributions inside and outside the cylindrical structure for $\gamma_1 = \frac{\rho_2}{\rho_1} = 1.9, \gamma_2 = \frac{\rho_3}{\rho_2} = 1.1145, \rho_1 = 23\ nm$, and $\epsilon_2 = -1.25$ (Figure 4). When the plane wave is normally incident onto theoptimized cylindrical structure with GST in its crystalline phase, the plane wave field distribution is not distorted at all by the presence of the structure (Figures 4(a) and 4(b)). On the other hand, when the plane wave is incident onto the optimized cylindrical structure with GST in its amorphous phase, the plane wave field distribution is severely distorted by the presence of the structure (Figure 4(c)), and the incident wave excites a dipole mode supported by the structure (Figure 4(d)). In Figure 2(a), we observe that in both regions I and III, which correspond to physical solutions, the dielectric permittivity of the inner shell ϵ_2 (Figure 1) is negative. In general, an electrically small cylindrical structure exhibits a dipolar scattered field due to the electric dipole moment excited by the incident wave. The polarization vector in a shell with negative dielectric constant being antiparallel tothose in other layers with positive dielectric constants may lead to the overall cancellation of dipole moment [14]. The dielectric permittivity of the inner shell has therefore to be negative to realize cloaking invisibility for our proposed optimized structure (Figure 1) with GST in its crystalline phase. A material with dielectric constant ϵ being in the range $-10 < \epsilon < -0.5$ at infrared and visible frequencies can be formed by embedding silver implants in a dielectric host with positive dielectric constant [85]. In addition, there exist natural low loss plasmonic materials with the required negative dielectric permittivity at mid-infrared wavelengths, such as highly doped InP [87]. Its permittivity and plasmonic properties can be tuned through the carrierconcentration [86]. In addition, we note that, even in the presence of loss in the inner shell, the contrast ratio is still close to unity [26]. As an example, we found that, for $\epsilon_2 = -1.25 + 0.1j$, the scattering cross section in the enhanced scattering regime is ~0.1, while the scattering cross section in the cloaking regime is ~8.0×10^{-7}.

2.2.2. Moderately Small Cylindrical Structures

When the physical size of the cylindrical structure increases, other Mie scattering order contributions to the overall scattering cross section also need to be taken into account [87]. Thus, Eq. (21) can no longer be used to determine the optimized geometrical and material parameters of the structure in Figure 1 for achieving switching between cloaking invisibility and enhanced scattering regimes, when GST is switched from its crystalline to its amorphous phase. In general, cloaking a moderately small structure through multiple scattering orders occurs when the numerators of the corresponding scattering coefficients (Eq. (13)) become zero at the same frequency. In addition, superscattering, which corresponds to the scattering cross section of a moderately small structure being significantly enhanced, occurs when overlap of at least two different scattering resonance modes is achieved at the same frequency [19, 20, 76]. In this subsection, the radius of the core layer is set equal to 480 nm which is approximately 10 times larger than the radius of the optimized electrically small three-layer cylindrical structure. In addition, the material of the inner shell is chosen to be TiO_2 with dielectric constant which can be approximated by the following Sellmeier dispersion equation [88, 89]

$$\epsilon_{TiO_2} = \epsilon_2 = 5.193 + \frac{0.244}{\lambda_0^2 - 0.0803} \tag{22}$$

where the wavelength λ_0 is in units of micrometers. Thus, the dielectric constant of TiO_2 at the operating wavelength of $\lambda_0 = 4\mu m$ is ~ 5.2. To realize switching between cloaking and superscattering regimes for the moderately small cylindrical structure of Figure 1, we now use Eqs. (14) and (15). More specifically, we optimize $\gamma_1 = \frac{\rho_2}{\rho_1}$ and $\gamma_2 = \frac{\rho_3}{\rho_2}$ (Figure 1) to make the NSCS of the structure [Eq. (16)] as large as possible when GST is in its csystalline phase, and as small as possible when GST is in its amorphous phase at the wavelength of $\lambda_0 = 4\ \mu m$.

In Figure 5, we show the NSCS σ_N of the proposed moderately small core-shell cylindrical structure for different scattering orders as a function of $\gamma_1 = \frac{\rho_2}{\rho_1}$ and $\gamma_2 = \frac{\rho_3}{\rho_2}$, when GST is in its crystalline phase (Figures 5(a)-5(d)), and its amorphous phase (Figures 5(e)-5(h)) at the wavelength of $\lambda_0 = 4\ \mu m$. For the range of parameters shown, the amplitudes of the higher order coefficients $|b_n|$ are negligible for $|n| > 3$. When the phase-change material GST is in its crystalline phase, the resonant scattering regime is broad in the $\gamma_1 - \gamma_2$ space for n = 0, but narrow for $|n| = 1, 2$ and 3 (Figures 5(a)-5(d)). Thus, it is very difficult to achieve overlapping of the dipole (n = 1), quadrupole (n = 2), and sextupole (n = 3) scattering modes. However, superscattering can be realized when there is overlap between the monopole (n = 0) and a higher order multipolar mode (dipole, quadrupole, or sextupole).

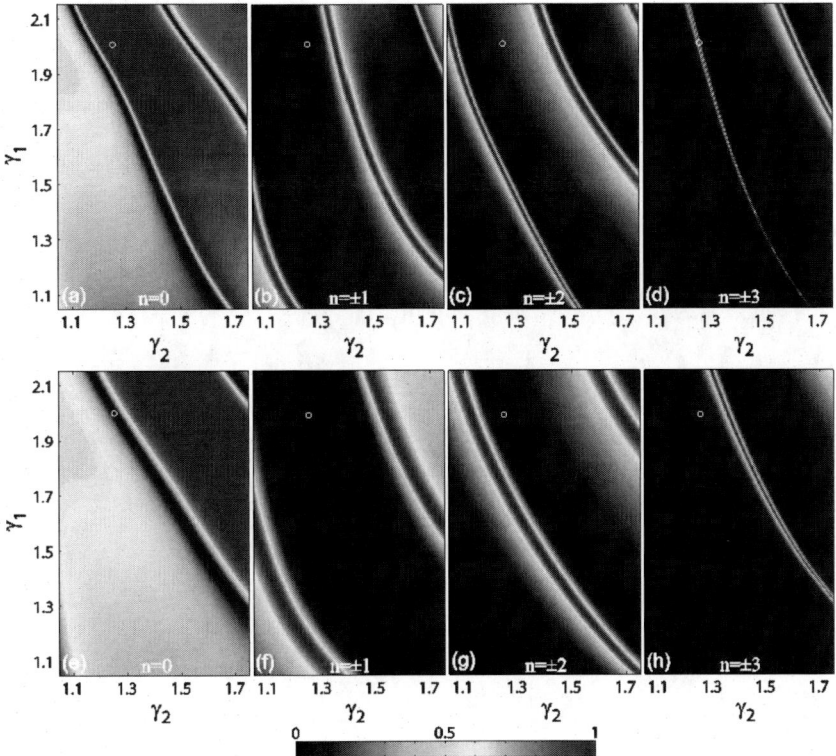

Figure 5. (a)-(d) The NSCS σ_N for cylindrical structures as in Figure 1 as a function of $\gamma_1 = \frac{\rho_2}{\rho_1}$ and $\gamma_2 = \frac{\rho_3}{\rho_2}$ for different scattering orders at the wavelength of $\lambda_0 = 4\mu m$ for GST in its crystalline phase. The material in the core layer is ZnO, and its radius is $\rho_1 = 480 nm$. The material in the inner shell is TiO_2. (e)-(h) Same as in (a)-(d) except that GST is in its amorphous phase. The three-layer core-shell structure is in the superscattering regime for GST in its crystalline phase, and in the cloaking regime for GST in its amorphous phase at the wavelength of $\lambda_0 = 4\mu m$, when $\gamma_1 = 1.99$ and $\gamma_2 = 1.26$ (white circle).

Thus, superscattering occurs when a higher order multipolar scattering mode is on resonance. When the phase-change material GST is in its amorphous phase, the near-zero scattering regime is narrow in the $\gamma_1 - \gamma_2$ space for n = 0, but quite broad for $|n| = 1, 2$ and 3 (Figures 5(e)-5(h)). For the range of parameters shown, the cloaking condition for the proposed moderately small structure can therefore be approximately reduced to $|b_0|^2 = 0$. In other words, cloaking occurs when the amplitude of the monopole scattering coefficient b_0 is zero. Switching between superscattering and cloaking regimes can be achieved when superscattering for GST in its crystalline phase and cloaking for GST in its amorphous phase occur at the same point in the $\gamma_1 - \gamma_2$ space. Figure 5 reveals that the optimized moderately small three-layer core-shell structure is in the superscattering regime for GST in its crystalline phase, and in the cloaking regime for GST in its amorphous phase at the wavelength of $\lambda_0 = 4\,\mu m$, when $\gamma_1 = 1.99$ and $\gamma_2 = 1.26$ (white circle in Figure 5). The radius of such a three-layer cylindrical structure is $\rho_3 = \rho_1 \gamma_1 \gamma_2 = 1023.5\,nm$, which is $\sim \frac{\lambda_0}{3.3}$.

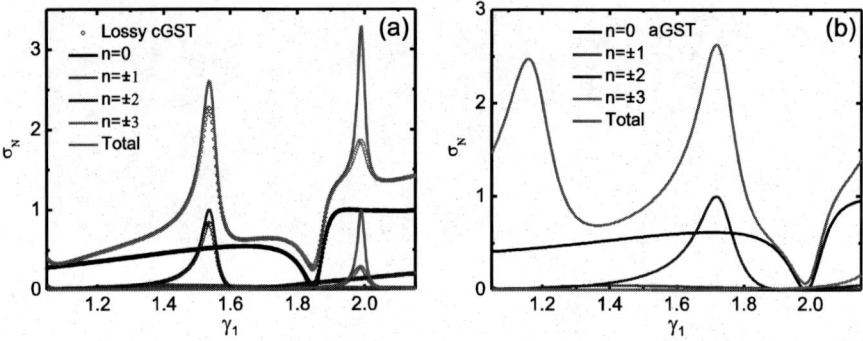

Figure 6. (a) The NSCS σ_N for cylindrical structures as in Figure 1 as a function of $\gamma_1 = \frac{\rho_2}{\rho_1}$ with $\gamma_2 = \frac{\rho_3}{\rho_2} = 1.26$ at the wavelength of $\lambda_0 = 4\mu m$ for GST in its crystalline phase. All other parameters are as in Figure 5(a). The solid lines and circles correspond to lossless and lossy cGST, respectively. (b) Same as in (a) except that GST is in its amorphous phase.

To further illustrate the superscattering property of the optimized structure, we show the NSCS σ_N for different scattering orders as a function of $\gamma_1 = \frac{\rho_2}{\rho_1}$ with $\gamma_2 = \frac{\rho_3}{\rho_2} = 1.26$ when GST is in its crystalline phase (Figure 11(a)). We observe that, when $\gamma_1 = 1.99$, the monopole, quadrupole and sextupole modes overlap very well. In the lossless cGST case, the overall σ_N, which is calculated with full-wave finite-element simulations, can reach 3.29 (Figure 6(a)). In the presence of loss, the overall σ_N is reduced to 1.85 (Figure 6(a)). Similarly, to further illustrate the cloaking property of the optimized structure, we show the NSCS σ_N for different scattering orders when GST is in its amorphous phase (Figure 6(b)). We observe that the overall σ_N can be drastically suppressed for $\gamma_1 = 1.99$, where the monopole mode contribution becomes zero, and the higher order multipolar mode contributions are very small. The corresponding overall σ_N can be as small as ~0.067 (Figure 6(b)). The contrast ratio τ between the supercattering and cloaking states is $\left|\frac{1.85-0.067}{1.85+0.067}\right|$ ~93%.

The switching between cloaking and superscattering regimes at the wavelength of $\lambda_0 = 4\ \mu m$ can be observed in the field distributions inside and outside the moderately small core-shell cylindrical structure of Figure 1 for $\gamma_1 = \frac{\rho_2}{\rho_1} = 1.99$, $\gamma_2 = \frac{\rho_3}{\rho_2} = 1.26$, and $\rho_1 = 480\ nm$ (Figure 7). Figure 7(a) shows the magnetic field profile for a bare ZnO core with radius of $\rho_1 = 480\ nm$. In this case, the overall σ_N is 0.77. The scattering is dominated by the monopole mode (Figure 7(b)). In Figures 7(c) and 7(d), we show the magnetic field distributions of the optimized three-layer cylindrical structure when GST is in its crystalline phase. When a plane wave is normally incident from the left, the core-shell cylindrical structure induces strong backscattering, and leaves a significant shadow in front of it, where the field strength is reduced (Figure 7(c)). The enhanced NSCS (~1.85) is 2.4 times larger than that of the bare ZnO core (~0.77). We observe that, when $\gamma_1 = 1.99$, the monopole, quadrupole, and sextupole modes overlap very well. The field distribution inside the cylindrical structure is a superposition of the monopole, quadrupole, and sextupole mode fields (Figure 7(d)), which is consistent with the results associated with superscattering in Figure 6(a). The monopole modal fields reside mostly in the lossless ZnO core. The strong overlap of the resonant sextupole modal fields with the lossy cGST outer shell leads to a greatly reduced scattering cross section compared to the lossless cGST case (Figure 6(a)). Interestingly, even though the material loss in cGST highly affects the sextupole mode, it does not affect much the dipole and quadrupole modes (Figure 6(a)). This is due to the weak overlap of the dipole and quadrupole modes with the lossy cGST shell. In addition, the resonant sextupole mode has a higher quality factor than that of the dipole and quadrupole modes (Figure 7(a)), which results in light being trapped in the structure for a longer duration. This in turn leads to higher power penalty [76, 90]. Similarly, in Figures 7(e) and 7(f) we show the magnetic field distribution of the optimized three-layer cylindrical structure when GST is in its amorphous phase. We observe that in this case there is hardly any scattering. The suppressed NSCS $\sigma_N(\sim 0.067)$ is 11.5 times smaller than that of the bare ZnO core(~ 0.77). Two different resonant modes can be observed in the core and shell regions (Figure 7(f)). These two excited resonant modes are out-of-phase and compensate each other in the far-field, resulting in scattering cancellation [16].

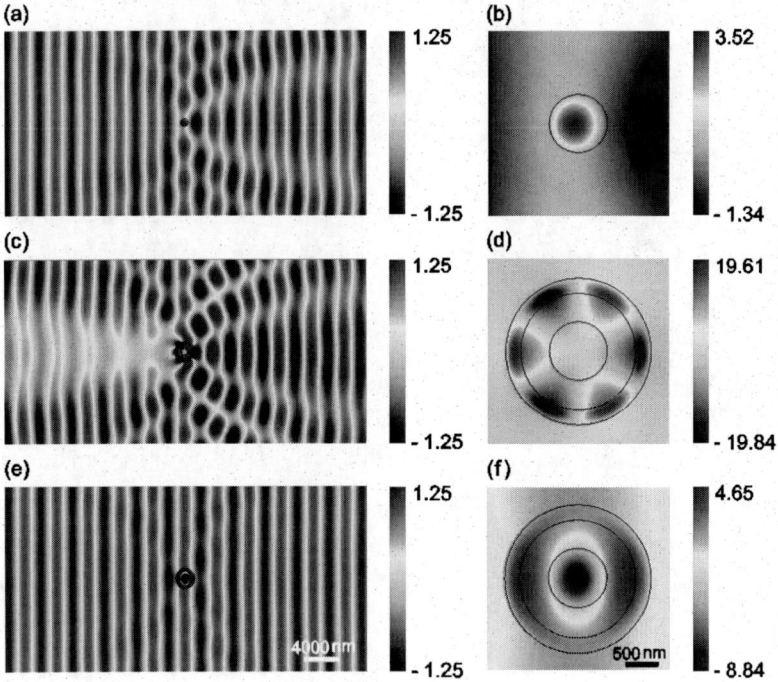

Figure 7. (a) and (b) Magnetic field amplitude profiles for a bare ZnO core with radius of $\rho_1 = 480\ nm$ at the wavelength of $\lambda_0 = 4\ \mu m$, when aplane wave is normally incident from the left. The fields are normalized with respect to the field amplitude of the incident planewave. (c) and (d) Magnetic field amplitude profiles for the optimized core-shell cylindrical structure of Figure 1 with GST in its crystalline phase and $\gamma_1 = \frac{\rho_2}{\rho_1}$. All other parameters are as in Figure 6(a). (e) and (f) Same as in (c) and (d) except that GST is in its amorphousphase.

As final remarks, we first investigated a two-layer cylindrical structure consisting of a core layer and a GST shell layer. We found that such a two-layer structure cannot achieve the same functionality as the three-layer structure. In other words, the inner shell between the core and the GST outer shell is necessary to achieve switching between cloaking and enhanced scattering regimes. This is due to the fact that for electrically small two-layer cylindrical structures, a shell with negative or near-zero permittivity is necessary to cloak a dielectric core based on scattering cancellation [14, 15]. Thus, an electrically small two-layer structure consisting of a GST shell and a dielectric core cannot be used to realize switching between the cloaking and resonant scattering regimes. We also note that layered core-shell cylindrical nanostructures can be fabricated by chemical vapor deposition and sputter coating [91]. Light scattered from an individual cylindrical nanostructure can be detected using dark-field microscopy [92]. In addition, switching between cloaking and superscattering regimes using phase-change materials could also be generalized to three-dimensional spherical nanostructures [20]. Our results could be potentially important for developing a new generation of dynamically reconfigurable subwavelength optical devices.

3. Switching of the Direction of Reflectionless Light Propagation at Exceptional Points in Non-PT-Symmetric Structures

3.1. Theory

Our proposed structure consists of a gain medium layer sandwiched between two GST layers (Figure 8). The optical properties of our proposed system can be described by the scattering matrix S defined by the following equation [54, 56, 57, 60, 70–72].

$$\begin{pmatrix} H_R^- \\ H_L^- \end{pmatrix} = S \begin{pmatrix} H_L^+ \\ H_R^+ \end{pmatrix} = \begin{pmatrix} t & r_b \\ r_f & t \end{pmatrix} \begin{pmatrix} H_L^+ \\ H_R^+ \end{pmatrix} \tag{22}$$

where H_L^+, and H_R^+ are the complex magnetic field amplitudes of the incoming waves at the left and right ports, respectively. Similarly, H_L^-, and H_R^- are the complex magnetic field amplitudes of the outgoing waves from the left and right ports, respectively. In addition, t is the complex transmission coefficient, while r_f, r_b are the complex reflection coefficients for uniform plane waves normally incident from the left (forward direction) and from the right (backward direction), respectively. In general, the matrix S is non-Hermitian, and its complex eigenvalues are $\lambda_S^\pm = t \pm \sqrt{r_f r_b}$. Its eigenstates, which are $\psi_\pm = (\sqrt{r_b}, \pm \sqrt{r_f})^T$ are not orthogonal. In our proposed three-layer optical system (Figure 8), by manipulating the elements of the scattering matrix, the two eigenvalues can be coalesced and form exceptional points. This leads to unidirectional reflectionless propagation in either the forward ($r_f = 0, r_b \neq 0$) or the backward ($r_b = 0, r_f \neq 0$) direction. We use the transfer matrix method to calculate t, r_f, and r_b for the three-layer structure of Figure 8 [93].

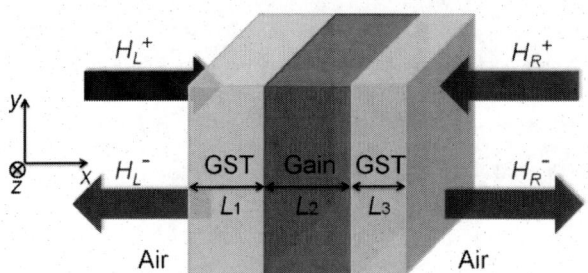

Figure 8. Schematic of a non-PT-symmetric three-layer structure composed of a gain medium layer sandwiched between two GST layers for switching of the direction of reflectionless light propagation at exceptional points.

When the system includes loss or gain, while the transmission coefficients in the forward and backward directions are the same, the reflection coefficients in the forward and backward directions are in general different. Such a system is analogous to open quantum systems which are characterized by complex non-Hermitian Hamiltonians [56-59, 62]. We will show that switching of the direction of reflectionless light propagation at exceptional points at the wavelength of 1.55µm can be achieved by proper tuning of the geometric and material parameters of the structure of Figure 8. More specifically, unidirectional reflectionlessness in the forward direction ($r_f = 0, r_b \neq 0$) can be switched to unidirectional reflectionlessness in the backward direction ($r_b = 0, r_f \neq 0$) by switching the phase-change material GST from its amorphous to its crystalline phase.

3.2. Results

In this section, we use a non-PT-symmetric three-layer structure, consisting of a gain medium layer sandwiched between two GST layers (Figure 8), to realize switching of the direction of reflectionless light propagation as discussed in the previous section. We found that both GST layers are critical for realizing the switching. More specifically, we first investigated a two-layer structure consisting of a gain medium layer and a single GST layer. We found that such a two-layer structure cannot be optimized to achieve the same functionality. We use the finite-difference frequency-domain (FDFD) method [94] to numerically calculate the reflection coefficients in the proposed structure. Perfectly matched layer (PML) absorbing boundary conditions are used at all boundaries of the simulation domain [95]. We also use the total-field-scattered-field (TFST) formulation to simulate the response of the structure to a normally incident plane wave input [96].

3.2.1. Switching of the Direction of Reflectionless Light Propagation

To realize switching of the direction of reflectionless light propagation, we use the transfer matrix method. More specifically, we optimize the thicknesses of all three layers L_1, L_2, and L_3 (Figure 8), as well as the imaginary part k of the refractive index $\tilde{n}_g$ of the gain material, to make the sum of the amplitude of the reflection coefficient in the forward direction r_f as close to zero as possible while of the amplitude of the reflection coefficient in the backward direction r_b as large as possible when GST is in its amorphous phase, and make the sum of the amplitude of the reflection coefficient in the backward direction $|r_b|$ as close to zero as possible while of the amplitude of the reflection coefficient in the forward direction r_f as large as possible when GST is in its crystalline phase at the optical communication wavelength of $\lambda_0 = 1.55\,\mu m$. We use experimental data for the frequency-dependent dielectric constant of amorphous GST (aGST) and crystalline GST (cGST) [28]. The refractive indices of aGST and cGST are $\tilde{n}_{aGST} = 4.7 + j0.2$ and $\tilde{n}_{cGST} = 7 + j2$, respectively, at $\lambda_0 = 1.55\,\mu m$ [28, 29]. In the presence of pumping, optical gain can be achieved using InGaAsP with InAs quantum dots (QDs) [97–101]. The real part of the refractive index for Inx Ga1 x AsyP1 y in the infrared is given by $n = 3.1 + 0.46y$ [97, 98], and is weakly dependent on the wavelength [97, 98]. Here, we assume that the real part of the refractive index of the gain material is 3.44. Since the transfer matrix method for the structure of Figure 8 is computationally very efficient, we are able to use an exhaustive search in the design parameter space (L_1, L_2, L_3, and k) to optimize the structure. Using this approach, we find that for $L_1 = 276$ nm, $L_2 = 84$ nm, $L_3 = 2$ nm, and $k = 0.67$ the reflection in the backward direction is almost zero ($R_b = |r_b|^2 \simeq 3.4820 \times 10^{-7}$) when GST is in its crystalline phase, and the reflection in the forward direction is also almost zero ($R_f = |r_f|^2 \simeq 7.1352 \times 10^{-5}$) when GST is in its amorphous phase at $\lambda_0 = 1.55\,\mu m$. In addition, the reflection in the forward direction when GST is in its crystalline phase is nonzero ($R_f = |r_f|^2 \simeq 0.743$), and the reflection in the backward direction when GST is in its amorphous phase is nonzero ($R_b = |r_b|^2 \simeq 0.5896$) as well. Thus, we conclude that for the structure of Figure 8 switching of the direction of reflectionless light propagation can be realized at the optical communication wavelength by switching the phase-change material GST between its amorphous and crystalline phases.

Figure 9(a) shows the reflection spectra for the structure of Figure 8 with GST in its crystalline phase calculated for both forward and backward directions using full-wave FDFD simulations for $L_1 = 276$ nm, $L_2 = 84$ nm,

$L_3 = 2$ nm, and $k = -0.67$. The FDFD results confirm that the optimized structure of Figure 8 with GST in its crystalline phase is unidirectional reflectionless at $f = 193.4$ THz ($\lambda_0 = 1.55\ \mu m$), since the reflection in the backward direction R_b is zero, while the reflection in the forward direction R_f is nonzero. We note that for all the structures investigated in this paper the transfer matrix method results are in excellent agreement with the FDFD results. In addition, the contrast ratio between the forward and backward reflection, defined as $\eta = \left|\frac{R_f - R_b}{R_f + R_b}\right|$ [59], as a function of frequency is shown in Figure 9(b). At the wavelength of $\lambda_0 = 1.55\ \mu m$, the contrast ratio almost reaches unity ($\eta \simeq 0.9999$). The unidirectional reflectionless propagation can be observed in the magnetic field distributions for $L_1 = 276$ nm, $L_2 = 84$ nm, $L_3 = 2$ nm, and $k = -0.67$ [Figures. 9(c)-9(f)]. When the plane wave is normally incident from the left (forward direction), the field decays significantly in the left GST layer due to strong absorption, and the incident and reflected waves in air form a strong interference pattern [Figures. 9(c) and 9(e)]. On the other hand, when the uniform plane wave is incident from the right (backward direction), the maximum field is at the interface between the left GST layer and the gain medium layer within the structure, and there is hardly any reflection due to the complete destructive interference between all the reflected waves at the right boundary of the structure [Figures. 9(d) and 9(f)]. Note that the bulk refractive index of GST that we used here is still valid for the ultrathin GST layer with 2 nm thickness [34, 39].

Figure 10(a) shows the reflection spectra for the structure of Figure 8 with GST in its amorphous phase calculated for both forward and backward directions using full-wave FDFD simulations for $L_1 = 276$ nm, $L_2 = 84$ nm, $L_3 = 2$ nm, and $k = -0.67$. The FDFD results also confirm that the optimized structure of Figure 8 with GST in its amorphous phase is unidirectional reflectionless at $f = 193.4$ THz ($\lambda_0 = 1.55\ \mu m$), since the reflection in the forward direction R_f is zero, while the reflection in the backward direction R_b is nonzero.

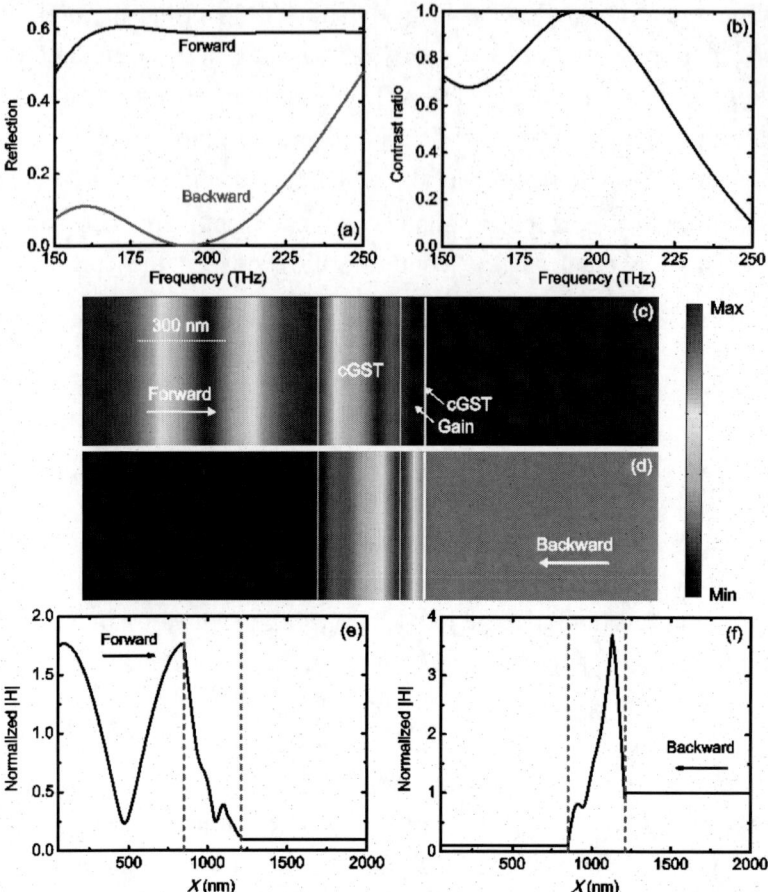

Figure 9. (a) Reflection spectra for the optimized structure of Figure 8 with GST in its crystalline phase calculated for normal incidence from both the forward and backward directions using FDFD. Results are shown for $L_1 = 276 nm$, $L_2 = 84 nm$, and $L_3 = 2 nm$. The gain medium is InGaAsP with InAs QDs ($\tilde{n}_g = n + jk = 3.44 - j0.67$). (b) Contrast ratio spectra for the optimized structure of Figure 8 with GST in its crystalline phase. All parameters are as in Figure 9(a). (c) and (d) Magnetic field amplitude profiles for the optimized structure of Figure 8 with GST in its crystalline phase at f = 193.4 THz($\lambda_0 = 1.55\ \mu m$), when a plane wave is normally incident from the left and right, respectively. All other parameters are as in Figure 9(a). (e) and (f) Magnetic field amplitude in the optimized structure of Figure 8 with GST in its crystalline phase at $f = 193.4$ THz ($\lambda_0 = 1.55\ \mu m$), normalized with respect to the field amplitude of the incident plane wave, when the light is incident from the left and right, respectively. The two vertical dashed lines indicate the left boundary of the left GST layer, and the right boundary of the right GST layer. All other parameters are as in Figure 9(a).

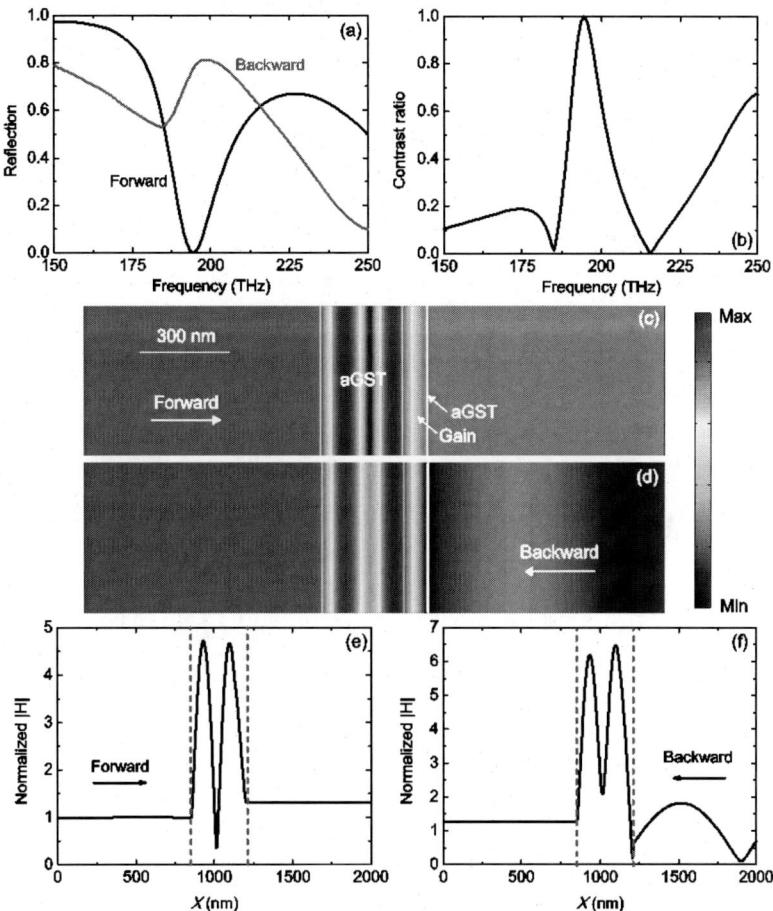

Figure 10. (a) Reflection spectra for the optimized structure of Figure 8 with GST in its amorphous phase calculated for normal incidence from both the forward and backward directions using FDFD. All parameters are as in Figure 9(a). (b) Contrast ratio spectra for the optimized structure of Figure 8 with GST in its amorphous phase. All parameters are as in Figure 9(a). (c) and (d) Magnetic field amplitude profiles for the optimized structure of Figure 8 with GST in its amorphous phase at $f = 193.4$ THz ($\lambda_0 = 1.55$ μm), when a plane wave is normally incident from the left and right, respectively. All other parameters are as in Figure 9(a). (e) and (f) Magnetic field amplitude in the optimized structure of Figure 8 with GST in its amorphous phase at $f = 193.4$ THz ($\lambda_0 = 1.55$ μ), normalized with respect to the field amplitude of theincident plane wave, when the light is incident from the left and right, respectively. The twovertical dashed lines indicate the left boundary of the left GST layer, and the right boundary of the right GST layer. All other parameters are as in Figure 9(a).

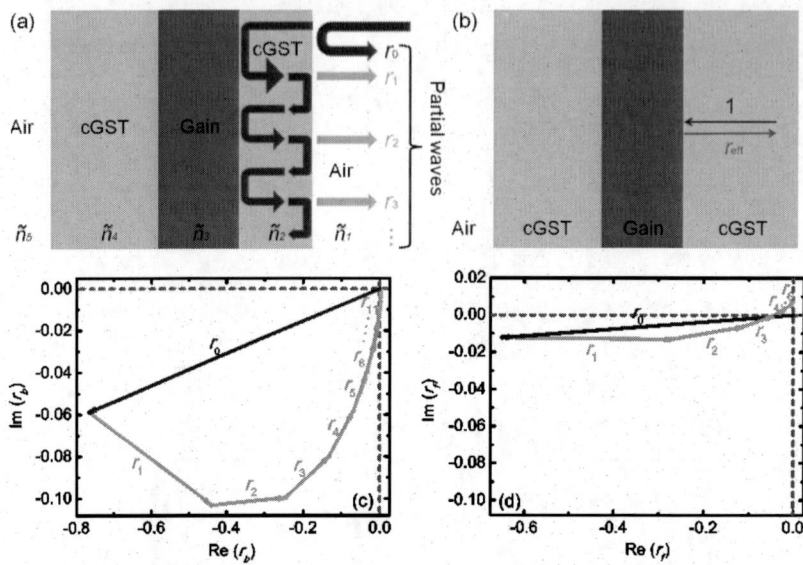

Figure 11. (a) The reflection process at normal incidence from a three-layer structure composed of a gain medium layer sandwiched between two GST layers showing the partial waves. GST is in its crystalline phase. (b) Schematic defining the reflection coefficient r_{eff} when a plane wave is normally incident on the boundary between GST and a two-layer structure composed of a gain medium layer and a GST layer above an air substrate. GST is in its crystalline phase. (c) Phasor diagram demonstrating that a zero-reflection condition is achievable via destructive interference for the optimized structure of Figure 8 with GST in its crystalline phase. A plane wave is incident from the right (backward direction) at $f = 193.4 THz$ ($\lambda_0 = 1.55 \mu m$). All other parameters are as in Figure 9(a). (d) Phasor diagram demonstrating that azero-reflection condition is achievable via destructive interference for the structure of Figure 8 with GST in its amorphous phase. A plane wave is incident from the left (forward direction) at $f = 193.4$ THz. All other parameters are as in Figure 9(a).

The contrast ratio between the forward and backward reflection as a function of frequency is shown in Figure 10(b). At the wavelength of $\lambda_0 = 1.55 \mu m$, the contrast ratio almost reaches unity($n \simeq 0.9999$). Similarly, the unidirectional reflectionless propagation can be observed in the magnetic field distributions for $L_1 = 276 nm$, $L_2 = 84\ nm$, $L_3 = 2\ nm$ and $k = -0.67$ [Figures 10(c)-10(f)]. When the plane wave is normally incident from the right (backward direction), the incident and reflected waves in air form a strong interference pattern [Figures 10(d) and 10(f)]. On the other hand, when the uniform plane wave is normally incident from the left (forward direction), there is hardly any reflection due to the complete destructive interference between all the reflected waves at the left boundary of the structure [Figures 10(c) and 10(e)].

As we saw above, in the optimized structure the thickness of the right GST layer L_3 is only 2 nm, which is much smaller than the wavelength of light $\lambda_0 = 1.55\ \mu m$. In general, there is hardly any phase shift in a few nm thick layer. However, if we remove this ultrathin layer from the optimized structure of Figure 8 with GST in its crystalline phase, the reflection in the backward direction increases significantly from 0 to 5.6%. This is due to the large extinction coefficient of cGST, as we will see below. In fact, if this ultrathin layer has an even higher extinction coefficient, the difference between the reflections from the structure with and without the ultrathin layer will be even larger. As an example, at the frequency of f = 250 THz, GST in its crystalline phase has a higher extinction coefficient (the refractive index is 6.45 + j3.0), compared to that at $f = 193.4$ THz ($\lambda_0 = 1.55\ \mu m$). Thus, at the frequency of $f = 250$ THz the reflection in the backward direction dramatically changes from 48.8% to 76.2%, if the 2 nm GST layer is removed from the optimized structure.

We further investigate the role of the ultrathin cGST layer with thickness L_3 in the reflection process from the three-layer structure of Figure 8. The ultrathin layer can be regarded as a cavity between air and the gain-cGST-air structure (Figure 11(a)). The total reflected wave in the backward direction can be calculated as the coherent sum of the partial wave reflected from the interface between air and cGST with reflection coefficient r_0, and those reflected from the 2 nm ultrathin cGST layer after one, two, three, ... roundtrips with reflection coefficients r_1, r_2, r_3, ... (Figure 11(a)). The total reflection coefficient in the backward direction r_b can then be obtained as [102]

$$r_b = \sum_{m=0}^{\infty} r_m = \frac{r_{12} + r_{eff} e^{2j\gamma L_3}}{1 + r_{12} r_{eff} e^{2j\gamma L_3}} \tag{23}$$

where $r_m = t_{12}r_{eff}^m r_{21}^{m-1}t_{21}e^{2j\gamma L_3}$ for $m > 0$. Here $r_{pq} = \frac{\tilde{n}_p - \tilde{n}_q}{\tilde{n}_p + \tilde{n}_q}$, $t_{pq} = \frac{2\tilde{n}_p}{\tilde{n}_p + \tilde{n}_q}$ are the Fresnel reflection and transmission coefficient when a plane wave is normally incident on medium q from medium p [93], $\tilde{n}_p$ is the complex refractive index of medium p, $r_0 = r_{12}$, and $\gamma = \frac{2\pi\tilde{n}_2}{\lambda_0}$. Here, $\tilde{n}_1 = \tilde{n}_5 = 1, \tilde{n}_2 = \tilde{n}_4 = \tilde{n}_{cGST}$, and $\tilde{n}_3 = \tilde{n}_g$, [Figure 11(a)]. Finally r_{eff} is the complex reflection coefficient when a plane wave is normally incident on the boundary between cGST and a two-layer structure composed of a gain medium layer and a cGST layer above an air substrat (Figure 11(b)).

In Figure 11(c), we show a phasor diagram, in which the reflected partial waves are plotted in the complex plane, to describe the reflection property in the backward direction of the optimized three-layer structure of Figure 8 with GST in its crystalline phase. The sum of the reflected partial waves, r_m, after m round trips in the ultrathin cGST layer destructively interferes with the first partial wave reflected from the interface between air and cGST, r_0. In the case of lossless dielectrics, phase accumulation occurs when light propagates through the layer. Light reflected at an interface between two lossless dielectrics can induce either π or 0 phase shifts [93]. However, when at least one of the dielectrics has a large extinction coefficient, the reflection coefficient is complex. The interface reflection can contribute a phase shift which is neither π nor 0, so that it is not necessary to achieve large phase accumulation via propagation through a thick dielectric layer as in the lossless case [102, 103]. Thus, due to the large extinction coefficient of cGST, a strong phase accumulation can be achieved by the ultrathin cGST layer, and the overall destructive interference leads to a zero reflection in the backward direction. The phasor path traces out a loop in the complex plane far from the horizontal axis, due to the large extinction coefficient of cGST, but returns to the origin since the destructive interference is complete (Figure 11(c)).

In Figure 11(d), we show a phasor diagram to describe the reflection property in the forward direction of the optimized three-layer structure of Figure 8 with GST in its amorphous phase, based on the same method used in Figure 11(c). All reflected partial waves destructively interfere almost completely, so that the phasor path returns to a point very close to the origin [Figure 11(d)]. In the case of lossless dielectrics, the phasor trajectory is along the horizontal axis [102, 103]. Even though the phasor trajectory shown in Figure 11(d) still traces out a loop in the complex plane due to loss in aGST, unlike the previous case of cGST [Figure 11(c)], the loop path is very close to the horizontal axis because of the small extinction coefficient of aGST. Since aGST has a small extinction coefficient, the reflection in the forward direction remains close to zero even if weremove the ultrathin right aGST layer. On the other hand, the left aGST layer has to be thick enough ($L_1 = 276\,nm$) to achieve sufficient phase shift for substantial destructive interference.

Figure 12(a) shows the reflection in the backward direction for the optimized structure with GST in its crystalline phase as a function of the real and imaginary parts of the complex refractive index of the gain medium. Similarly, Figure 12(b) shows the reflection in the forward direction for the optimized structure with GST in its amorphous phase. We observe that the material of the middle layer has to exhibit gain ($k < 0$) in order to realize zero reflection in the backwarddirection ($R_b = 0$) for GST in its crystalline phase, and zero reflection in the forward direction ($R_f = 0$) for GST in its amorphous phase. This is because, in the absence of gain, each reflected partial wave would be highly attenuated after light propagates several round trips through thelossy GST layers, and the sum of reflection coefficientsr_1, r_2, r_3, ... would not cancel out the primary reflection coefficient r_0. Thus, gain material has to be used in our optimized structure in order to compensate the material loss in the GST layers and achieve complete destructive interference. Note that the imaginary part of the refractive index of the gain material used in theoptimized structure is 0.67, which corresponds to a gain coefficient of $g \approx 47580\ cm^{-1}$[104]. Also note that the near zero reflection regime is broad in the $n - k$ space [Figures 12(a) and 12(b)], which relaxes the requirements for realization of our proposed structures. In recent related experiments on unidirectional reflectionless structures the measured contrast ratio was 70% [59]. We found that our proposed structures can have contrast ratios of at least 70% using a significantly lower material gain of ~32000cm^{-1}. Even though achieving the material gain required for our design is challenging, it could be realized with QDs. QDs exhibit exceptionally large material gain [105, 106]. In addition, structures with ultra-high-density QDs and therefore very large QDs volume ratio have been recently reported [107, 108]. Such ultra-high-density QD structures could achieve the material gain required for our design. We finally note that the required material gain coefficient can be greatly reduced if the thickness of the gain medium layer is increased. Thus, we found that for optimized structures with gain medium layer thicknesses of 543 nm and 1220 nm the required material gain coefficients are 9090 cm^{-1} and 4070 cm^{-1}, respectively.

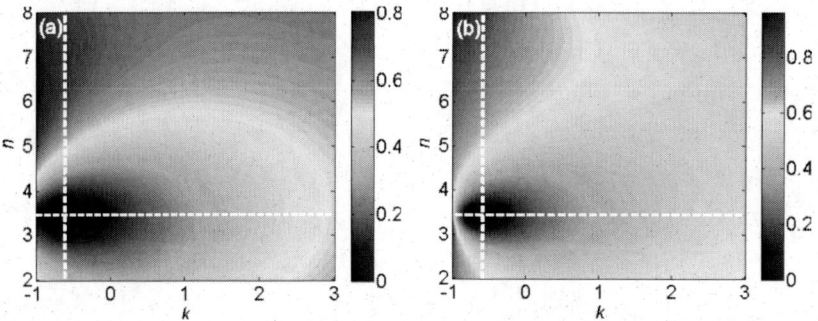

Figure 12. (a) Calculated reflection for the structure of Figure 8 with GST in its crystalline phase as a function of the real and imaginary parts, n and k, of the refractive index of the gain material. Results are shown for normal incidence from the backward direction at f = 193.4 THz ($\lambda_0 = 1.55\ \mu m$). All other parameters are as in Figure 9(a). The dashed lines correspond to the n and k of the optimized structure ($\tilde{n}_g = n + jk = 3.44 - j0.67$). (b) Same as in (a) except that GST is in its amorphous phase and results are shown for normal incidence from the forward direction.

3.2.2. Exceptional Points and Their Topological Properties

In the presence of loss or gain, the optical system of Figure 8 is analogous to open quantum systems which are characterized by complex non-Hermitian Hamiltonians [6, 57–60], and there is a close analogy between optical scattering matrices and Hamiltonian matrices [6, 64, 65, 69, 70,]. When the optical scattering matrix eigenvalues coalesce into a single eigenvalue, and the eigenstates coalesce into a single eigenstate, the system exhibits an optical exceptional point [6, 54, 59, 60]. This leads to unidirectional reflectionless propagation in either the forward or the backward direction. In the non-PT-symmetric three-layer system of Figure 8 with $L_1 = 276\, nm$, $L_2 = 84\, nm$, $L_3 = 2\, nm$, and $k = -0.67$ the existence of exceptional points for GST in both its amorphous and crystalline phases, leads to unidirectional reflectionless propagation in the forward direction $r_f = 0, r_b \neq 0$ for GST in its amorphous phase, and in the backward direction $r_b = 0, r_f \neq 0$ for GST in its crystalline phase at $f = 193.4$ THz ($\lambda_0 = 1.55\, \mu m$).

In Figure 13(a), we observe that the phase of the reflection coefficient in the backward direction r_b for the optimized three-layer structure of Figure 8 with GST in its crystalline phase undergoes an abrupt π jump, when the frequency is crossing over the exceptional point, which actually resembles the phase transition from the PT-symmetric phase to the PT-broken phase in optical PT-symmetric systems [6, 56, 59, 60]. In contrast, the phase of the reflection coefficient in the forward direction r_f varies smoothly with frequency [6, 56]. Similarly, we observe that the phase of the reflection coefficient in the forward direction r_f for the optimized three-layer structure of Figure 8 with GST in its amorphous phase undergoes an abrupt π jump, while the phase of the reflection coefficient in the backward direction r_b varies smoothly with frequency (Figure 13(b)).

To gain further insight into the properties of the exceptional point for the non-PT-symmetric system of Figure 8, we also investigate the topological structure of this exceptional point. In open quantum systems, the eigenvalues of a 2×2 non-Hermitian matrix can be expanded in Puiseux series around exceptional points [41]. In the vicinity of an exceptional point, the two complex eigenvalues are almost degenerate and can be approximated as [41]

$$E_{\pm}(x) = E_{EP} \pm \alpha \sqrt{x - x_{EP}} \qquad (24)$$

where $E_{\pm}(x)$ and x_{EP} are the eigenvalues and the corresponding exceptional point for the 2×2 non-Hermitian matrix in parameter x space, respectively, and α is complex. When $x - x_{EP} < 0$, the coefficient α has to be multiplied by a factor of $j = e^{j\frac{\pi}{2}}$, compared to the case of $x - x_{EP} > 0$. Thus, as the parameter x increases in the interval $[x_{EP} - \delta, x_{EP} + \delta]$, where δ is sufficiently small, the two eigenvalues for $x - x_{EP} < 0$ should move along a line which is perpendicular to that for $x - x_{EP} > 0$ [41, 42].

Figure 14(a) shows the trajectories of the two eigenvalues $\lambda_s^{\pm}$ of the scattering matrix S [Eq. (22)] for the optimized structure of Figure 8 with GST in its crystalline phase as the thickness L_2 approaches the exceptional point ($L_2 = 84\ nm$). We observe that for $L_2 < 84\ nm$ the two eigenvalues first move along the blue lines as L_2 increases, and coalesce at the exceptional point for $L_2 = 84\ nm$. As L_2 further increases ($L_2 > 84\ nm$), the two degenerate eigenvalues are split again and move along the red lines which are indeed perpendicular to the blue ones around the exceptional point. Figure 14(b) shows the trajectories of the two eigenvalues $\lambda_s^{\pm}$ for the optimized structure of Figure 8 with GST in its amorphous phase. Similarly, we observe that the two eigenvalues coalesce at the exceptional point, and the angle between the in and out trajectories around the exceptional point is ~90° in the complex eigenvalue plane, when L_2 sweeps over the exceptional point.

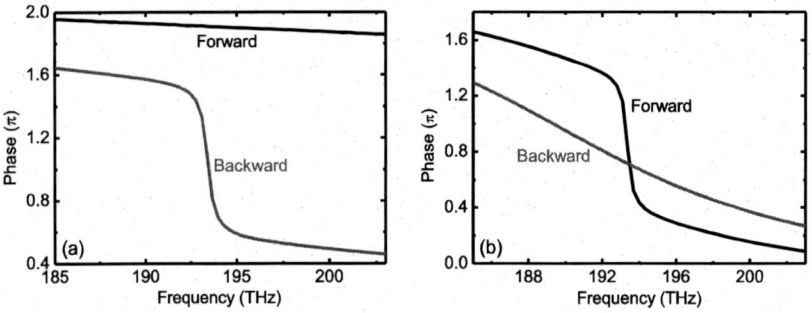

Figure 13. (a) Phase spectra of the reflection coefficients in the forward (r_f, black) and backward (r_b, red) directions for the structure of Figure 8 with GST in its crystalline phase. All parameters are as in Figure 9(a). (b) Same as in (a) except that GST is in its amorphous phase.

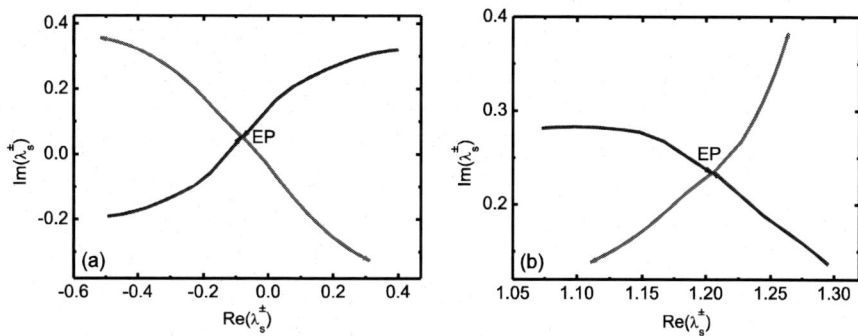

Figure 14. (a) Schematic representation of the coalescence of the two eigenvalues $\lambda_s^{\pm}$ of the scattering matrix S [Eq. (22)], as the thickness L_2 is varied for the structure of Figure 8 with GST in its crystalline phase. The red lines correspond to the eigenvalues for $L_2 \geqq 84\ nm$, the blue lines correspond to the eigenvalues for $L_2 < 84\ nm$, and the meeting-point of the arrows corresponds to the exceptional point. Results are shown for $f = 193.4\ \text{THz}(\lambda_0 = 1.55\ \mu m)$. All other parameters are as in Figure 9(a). (b) Same as in (a) except that GST is in its amorphous phase.

Another unique topological property of exceptional points is that, when encircling an exceptional point in parameter space, the eigenvalues will be interchanged after a complete loop, because of the square root behavior of the singularity [41, 42, 109-112]. In our non-PT-symmetric system (Figure 8), we consider the parameter space of the real and imaginary parts, n and k, of the complex refractive index of the gain material. The exceptional point is encircled by a circular loop with a radius of R in the n-k plane and is located at the center of this circle ($\tilde{n}_{g,EP} = n_{EP} + jk_{EP}$) [Figure 15(a)]. The radius R is chosen as 0.05 to ensure that only the exceptional point that we obtained by optimizing the structure is embedded inside this circle. We vary the refractive index of the gain material in the counterclockwise direction along the circular loop from the initial position A with $\tilde{n}_g = 3.44 - j0.67$ [Figure 15(a)]. That is, e vary the refractive index $\tilde{n}_g$ such that

$$\tilde{n}_g - \tilde{n}_{g,EP} = Re^{j\phi} \tag{25}$$

where ϕ is adiabatically varied from $\phi = 0$ to $\phi = 2\pi$.

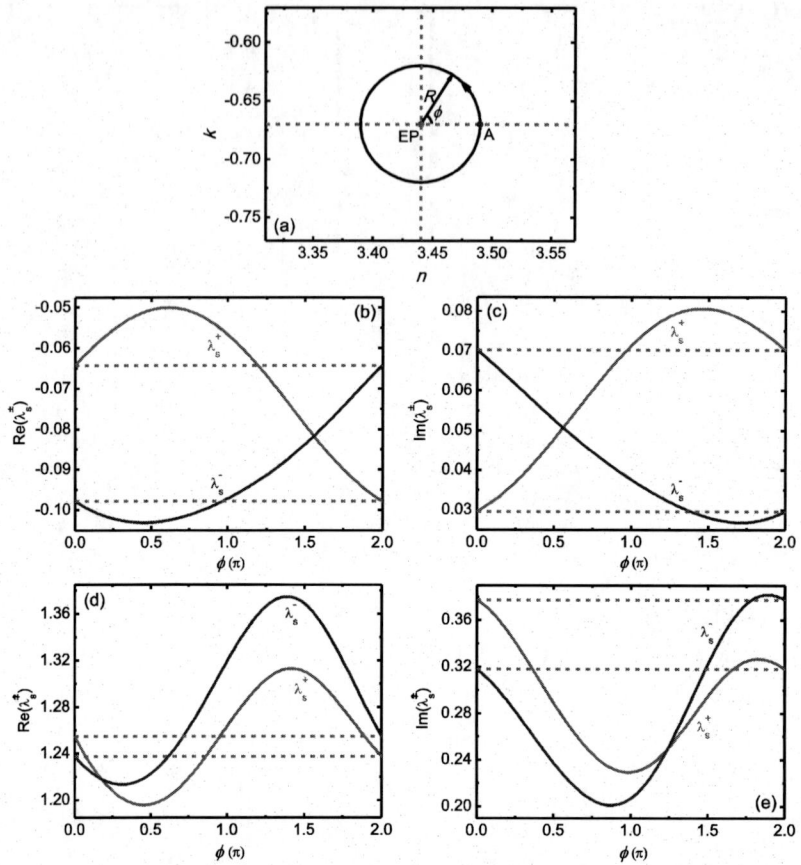

Figure 15. (a) A circular loop in the parameter space of the real and imaginary parts, n and k, of the complex refractive index of the gain material. The circle is centered at the exceptional point (red dot with $\tilde{n}_g = 3.44 - j0.67$), and its radius R is set to be 0.05. The blue dot represents the starting position of the loop (point A with $\tilde{n}_g = 3.44 - j0.67$). (b) and (c) The trajectories of the real and imaginary parts of the eigenvalues $\lambda_S^\pm$ of the scattering matrix S [Eq. (22)] for the structure of Figure 8, as the path of the complex refractive index of the gain material traces the circular loop of Figure 15(a) in the counterclockwise orientation. Results are shown for GST in its crystalline phase and f = 193.4 THz($\lambda_0 = 1.55\ \mu m$). All other parameters are as in Figure 9(a). (d) and (e). Same as in (b) and (c) except that GST is in its amorphous phase.

Figures 15(b) and 15(c) show the trajectories of the real and imaginary parts of the eigenvalues $\lambda_S^\pm$ of the scattering matrix S [Eq. (22)] for the structure of Figure 8 with GST in its crystalline phase, as the path of the complex refractive index of the gain material traces the circular loop of

Figure 15(a), which encircles the exceptional point, in the counterclockwise orientation. We observe that the two complex eigenvalues are interchanged after one loop of 2π.

The interchange of the two eigenvalues further confirms the existence of the exceptional point in the non-PT-symmetric system of Figure 8. The eigenvalues will return to their original values after two loops with the same orientation, which indicates that the exceptional point is a second order branch point for the eigenvalues [41, 42]. Figures 15(d) and 15(e) also show that the two complex eigenvalues are interchanged after one loop of 2π for the structure of Figure 8 with GST in its amorphous phase. In contrast, if there is no exceptional point in the closed loop, the eigenvalues will return to themselves at the end of the loop [41, 110]. Note that the switching of the eigenvalues implies the existence of an exceptional point inside the loop. Due to this unique topological structure of exceptional points, the switching of the eigenvalues indicates the existence of the exceptional point inside the loop without the need to locate the exact parameters at which the exceptional point occurs [41, 110, 112].

As final remarks, we first investigated a two-layer cylindrical structure consisting of a core layer and a GST shell layer. We found that such a two-layer structure cannot achieve the same functionality as the three-layer structure. In other words, the inner shell between the core and the GST outer shell is necessary to achieve switching between cloaking and enhanced scattering regimes. This is due to the fact that for electrically small two-layer cylindrical structures, a shell with negative or near-zero permittivity is necessary to cloak a dielectric core based on scattering cancellation [14, 15]. Thus, an electrically small two-layer structure consisting of a GST shell and a dielectric core cannot be used to realize switching between the cloaking and resonant scattering regimes. We also note that layered core-shell cylindrical nanostructures can be fabricated by chemical vapor deposition and sputter coating [91]. Light scattered from an individual cylindrical nanostructure can be detected using dark-field microscopy [92]. In addition, switching between cloaking and superscattering regimes using phase-change materials could also be generalized to three-dimensional spherical nanostructures [20]. Our results could be potentially important for developing a new generation of dynamically reconfigurable subwavelength optical devices.

As final remarks, for experimental realization, an ultrathin thermal barrier layer can be used next to a GST layer to confine heat in the GST film. Such a thermal barrier layer keeps surrounding materials isolated from heating and protects them from harmful interaction with the GST film, when GST is switched between its two phases by optically, electrically or thermally provided heat stimuli [29, 34, 39]. For example, Ta_2O_5, Al_2O_3 and Si_3N_4 are materials which can be chosen for the thermal barrier due to their low thermal conductivities [29, 34, 39]. We found that the addition of such a thin thermal barrier layer does not affect the functionality of the proposed structure. In addition, even though the structure was designed assuming normally incident light, we found that the contrast ratio between the forward and backward reflection is large in a wide angular range for both TE and TM polarization, and for GST in both its amorphous and crystalline phases. Our results could be potentially important for developing a new generation of compact active free-space optical devices. It is noteworthy that the concept of combining gain and phase-change materials for switching of the direction of reflectionless light propagation could also be applied to nanoplasmonic waveguide-cavity systems [6, 65], which could lead to implementations in integrated optical chips.

REFERENCES

[1]   Novotny, L. and Hecht, B. (2006). *Principles of Nano-Optics* (Cambridge University).

[2]   Barnes, W. L., Dereux, A. and Ebbesen, T. W. (2003). Surface plasmon subwave length optics. *Nature*, 424: 824–830.

[3]   Lal, S., Link, S. and Halas, N. J. (2007). Nano-optics from sensing to wave guiding. *Nat. Photonics*, 1: 641–648.

[4]   Stockman, M. I. (2011). Nanoplasmonics: past, present, and glimpse into future. *Opt. Express*, 19: 22029–22106.

[5] Yang, Z., Jiang, R., Zhuo, X., Xie, Y., Wang, J. and Lin. H. (2017). Dielectric nanoresonators for light manipulation. *Phys. Rep.*, 701: 1–50.

[6] Huang, Y., Veronis, G. and Min, C. (2015). Unidirectional reflectionless propagation in plasmonic waveguide-cavity systems at exceptional points. *Opt. Express*, 23: 29882–29895.

[7] Hirsch, L. R., Stafford, R. J., Bankson, J. A., Sershen, S. R., Rivera, B., Price, R. E., Hazle, J. D., Halas, N. J. and West, J. L. (2003). Nanoshell-mediated near-infrared thermal therapy of tumors under magnetic resonance guidance. *Proc. Natl. Acad. Sci, USA*, 100: 13549–13543.

[8] Jackson J. B. and Halas, N. J. (2004). Surface-enhanced Raman scattering on tunable plasmonic nanoparticle substrates. *Proc. Natl. Acad. Sci. USA*, 101: 17930–17935.

[9] Atwater, H. A. and Polman, A. (2010). Plasmonics for improved photovoltaic devices. *Nat. Mater.*, 9: 205–213.

[10] Alu, A. and Engheta, N. (2009). Cloaking a sensor. *Phys. Rev. Lett.*, 102: 233901.

[11] Fan, P., Chettiar, U. K., Cao, L., Afshinmanesh, F., Engheta, N. and Brongersma, M. L. (2012). An invisible metal-semiconductor photodetector. *Nat. Photonics*, 6: 380–385.

[12] Alu, A. and Engheta, N. (2010). Cloaked near-field scanning optical microscope tip for noninvasive near-field imaging. *Phys. Rev. Lett.*, 105: 263906.

[13] Mirzaei, A., Miroshnichenko, A. E., Shadrivov, I. V. and Kivshar. Y. S. (2015). Optical metacages. *Phys. Rev. Lett.*, 115: 215501.

[14] Alu, A. and Engheta, N. (2005). Achieving transparency with plasmonic and metamaterial coatings. *Phys. Rev, E*, 72: 016623.

[15] Chen, P., Soric, J. and Alu, A. (2012). Invisibility and cloaking based on scattering cancellation. *Adv. Mater.*, 24: OP281–OP304.

[16] Mirzaei, A., Miroshnichenko, A. E., Shadrivov, I. V. and Kivshar, Y. S. (2015). All-dielectric multilayer cylindrical structures for invisibility cloaking. *Sci. Rep.*, 5: 9574.

[17] Muhlig, S., Farhat, M., Rockstuhl, C. and Lederer, F. (2011). Cloaking dielectric spherical objects by a shell of metallic nanoparticles. *Phys. Rev, B*, 83: 195116.

[18] Monti, A., Bilotti, F. and Toscano, A. (2011). Optical cloaking of cylindrical objects by using covers made of core-shell nanoparticles. *Opt. Lett*, 36: 4479–4481.

[19] Ruan, Z. and Fan, S. (2010). Superscattering of light from subwavelength nanostructures. *Phys. Rev. Lett.*, 105: 013901 (2010).

[20] Ruan, Z. and Fan, S. (2011). Design of subwavelength superscattering nanospheres. *Appl. Phys. Lett.*, 98: 043101.

[21] Mirzaei, A., Miroshnichenko, A. E., Shadrivov, I. V. and Kivshar, Y. S. (2014). Superscattering of light optimized by a genetic algorithm. *Appl. Phys. Lett.*, 105: 011109.

[22] Huang, Y. and Gao, L. (2014). Superscattering of light from core-shell nonlocal plasmonic nanoparticles. *J. Phys. Chem. C*, 118: 30170–30178.

[23] Li, X., Tan, Q., Bai, B. and Jin, G. (2011). Experimental demonstration of tunable directional excitation of surface plasmon polaritons with a subwavelength metallic double slit. *Appl. Phys. Lett.*, 98: 251109.

[24] Lin, J., Mueller, J. B., Wang, Q., Yuan, G., Antoniou, N., Yuan, X. and Capasso, F. (2013). Polarization-controlled tunable directional coupling of surface plasmonpolaritons. *Science*, 340: 331–334.

[25] Kim, S., Yun, H., Park, K., Hong, J., Yun, J., Lee, K., Kim, J., Jeong, S., Mun, S., Sung, J., Lee, Y. and Lee, B. (2017). Active directional switching of surface plasmonpolaritons using a phase transition material. *Sci. Rep.*, 7: 43723.

[26] Argyropoulos, C., Chen, P., Monticone, F., Aguanno, G. D. and Alu, A. (2012). Nonlinear plasmonic cloaks to realize giant all-optical scattering switching. *Phys. Rev. Lett.*, 108: 263905.

[27] Chen, X., Sandoghdar, V. and Agio, M. (2013). Coherent interaction of light with a metallic structure coupled to a single quantum emitter: from superabsorption to cloaking. *Phys. Rev. Lett.*, 110: 153605.

[28] Shportko, K., Kremers, S., Woda, M., Lencer, D., Robertson, J. and Wuttig, M. (2008). Resonant bonding in crystalline phase-change materials. *Nat. Mater.*, 7: 653–658.

[29] Rude, M., Mkhitaryan, V., Cetin, A. E., Miller, T. A., Carrilero, A., Wall, S., Abajo, F. J., Altug, H. and Pruneri, V. (2016). Ultrafast and broadband tuning of resonant optical nanostructures using phase-change materials. *Adv. Optical Mater.*, 4: 1060–1066.

[30] Loke, D., Lee, T. H., Wang, W. J., Shi, L. P. Zhao, R., Yeo, Y. C., Chong, C. T. and Elliott, S. R. (2012). Breaking the speed limits of phase-change memory. *Science*, 336: 1556–1569.

[31] Hira, T., Homma, T., chiyama, T. U., Kuwamura, K., Kihara, Y. and Saiki, T. (2015). All-optical switching of localized surface plasmon resonance in single gold nanosandwich using GeSbTe film as an active medium. *Appl. Phys. Lett.*, 106: 031105.

[32] Wuttig M. and Yamada, N. (2007). Phase-change materials for rewriteable data storage. *Nat. Mater.*, 6: 824–832.

[33] Lankhorst, M. H. R., Ketelaarsand, B. W. and Wolters, R. A. M. (2005). Low-cost and nanoscale non-volatile memory concept for future silicon chips. *Nat. Mater.*, 4: 347–352.

[34] Schlich, F. F., Zalden, P., Lindenberg, A. M. and Spolenak, R. (2015). Color switching with enhanced optical contrast in ultrathin phase change materials and semiconductors induced by femtosecond laser pulses. *ACS Photonics*, 2: 178–182.

[35] Rude, M., Simpson, R. E., Quidant, R., Pruneri, V. and Renger, J. (2015). Active control of surface plasmon waveguides with a phase change material. *ACS Photonics*, 2: 669–674.

[36] Wang, Q., Rogers, E. T. F., Gholipour, B., Wang, C., Yuan, G., Teng, J., Zheludev, N. I. (2016). Optically reconfigurable metasurfaces and photonic devices based on phase change materials. *Nat. Photonics*, 10: 60–65.

[37] Li, P., Yang, X., Mab, T. W. W., Hanss, J., Lewin, M., Michel, A. U., Wuttig, M. and Taubner, T. (2016). Reversible optical switching of highly confined phonon-polaritons with an ultrathin phase-change material. *Nat. Mater.*, 15: 870–876.

[38] Mkhitaryan, V. K., Ghosh, D. S., Rude, M., Canet-Ferrer, J., Maniyara, R. A., Gopalan, K. K. and Pruneri, V. (2017). Tunable complete optical absorption in multilayer structures including $Ge_2Sb_2Te_5$ without lithographic patterns. *Adv. Optical Mater.*, 5: 1600452.

[39] Yoo, S., Gwon, T., Eom, T., Kim, S. and Hwang, C. (2016). Multicolor changeable optical coating by adopting multiple layers of ultrathin phase change material film. *ACS Photonics*, 3: 1265–1270.

[40] Huang, Y., Shen, Y., Min, C. and Veronis, G. (2017). Switching of the direction of reflectionless light propagation at exceptional points in non-PT-symmetric structures using phase-change materials. *Opt. Express*, 25: 27283–27297.

[41] Moiseyev, M. (2011). *Non-Hermitian Quantum Mechanics* (Cambridge University).

[42] Heiss, W. D. (2004). Exceptional points of non-Hermitian operators. *J. Phys. Math. Gen.*, 37: 2455–2464.

[43] Heiss, W. D. (2000). Repulsion of resonance states and exceptional points. *Phys. Rev.*, E61: 929–932.

[44] Lupu, A., Benisty, H. and Degiron, A. (2013). Switching using PT-symmetry in plasmonic systems: positive role of the losses. *Opt. Express*, 21: 21651–21668.

[45] Lupu, A., Benisty, H. and Degiron, A. (2014). Using optical PT-symmetry for switching applications. *Photon. Nanostructures*, 12: 305–311.

[46] Achilleos, V., Theocharis, G., Richoux, O. and Pagneux, V. (2017). Non-Hermitian acoustic metamaterials: Role of exceptional points in sound absorption. *Phys. Rev, B* 95: 144303.

[47] Yu, S., Park, H., Piao, X., Min, B. and Park, N. (2016). Low-dimensional optical chirality in complex potentials. *Optica*, 3: 1025–1032.

[48] Berry, M. V. (2004). Physics of nonhermitian degeneracies. *Czech. J. Phy.*, 54: 1039–1047.

[49] Ryu, J. W., Son, W. S., Hwang, D. U., Lee, S. Y. and Kim, S. W. (2015). Exceptional points in coupled dissipative dynamical systems. *Nano Lett.*, 91: 052910.

[50] Kirillov, O. N. (2013). Exceptional and diabolical points in stability questions. *Fortschr. Phys.*, 61: 205–224.

[51] Stehmann, T., Heiss, W. S. and Scholtz. (2004). Observation of exceptional points in electronic circuits. *J. Phys. Math. Gen.*, 37: 7813–7819.

[52] Heiss, W. D. (2012). The physics of exceptional points. *J. Phys. A*, 45: 444016.

[53] Bender, C. M. and Boettcher, S. (1998). Real spectra in non-Hermitian Hamiltonians having PT–symmetry. *Phys. Rev. Lett.*, 80: 5243.

[54] Lin, Z., Ramezani, H., Eichelkraut, T., Kottos, T., Cao, H. and Christodoulides, D. N. (2011). Unidirectional invisibility induced by PT-symmetric periodic structures. *Phys. Rev. Lett*, 106: 213901.

[55] Regensburger, A., Bersch, C., Miri, M. A., Onishchukov, G., Christodoulides, D. N. and Peschel, U. (2012). Parity-time synthetic photonic lattices. *Nature*, 488: 167–171.

[56] Ge, L., Chong, Y. D. and Stone, A. D. (2012). Conservation relations and anisotropic transmission resonances in one-dimensional PT-symmetric photonic heterostructures. *Phys. Rev*, A 85: 023802.

[57] Huang, Y., Shen, Y. Min, C., Fan, S. and Veronis, G. (2017). Unidirectional reflectionless light propagation at exceptional points. *Nanophotonics*, 6: 977–996.

[58] Yin, X. and Zhang, X. (2013). Unidirectional light propagation at exceptional points. *Nat. Mate.*, 12: 175–177.

[59] Feng, L., Xu, Y. L. W., Fegadolli, S., Lu, M. H., Oliveira, J. E. B., Almeida, V. R., Chen, Y. F. and Scherer, A. (2013). Experimental demonstration of a unidirectional reflectionless parity-time metamaterial at optical frequencies. *Nat. Mate.*, 12: 108–113.

[60] Feng, L., Zhu, X., Yang, S., Zhu, H., Zhang, P., Yin, X., Wang, Y. and Zhang, X. (2014). Demonstration of a large-scale optical exceptional point structure. *Opt. Express*, 22: 1760–1767.

[61] Shen, Y., Deng, X. H. and Chen, L. (2014). Unidirectional invisibility in a two-layer non-PT-symmetric slab. *Opt. Express*, 22: 19440–19447.

[62] Peng, B., Ozdemir, S. K., Liertzer, M., Chen, W., Johannes, K., Yilmaz, H., Wiersig, J., Rotter. S, and Yang, L. (2016). Chiral modes and directional lasing at exceptional points. *Proc. Natl. Acad. Sci. USA*, 113: 6845–6850.

[63] Jia, Y., Yan, Y., Kesava, S. V., Gomez, Z. D. and Giebink, N. C. (2015). Passive parity-time symmetry in organic thin film waveguides. *ACS Photonics*, 2: 319–325.

[64] Horsley, S. A. R., Artoni, M. and La Rocca, G. C. (2015). Spatial Kramer-Kronig relations and the reflection of waves. *Nat. Photonics*, 9: 436–439.

[65] Huang, Y., Min, C. and Veronis, G. (2016). Broadband near total light absorption in non-PT-symmetric waveguide-cavity systems. *Opt. Express*, 24: 22219–22231.

[66] Yang, E., Lu, Y., Wang, Y., Dai, Y. and Wang, P. (2016). Unidirectional reflectionless phenomenon in periodic ternary layered material. *Opt. Express*, 24: 14311–14321.

[67] Yu, S., Piao, X. and Park, N. (2016). Acceleration toward polarization singularity inspired by relativistic EB drift. *Sci. Rep.*, 6: 37754.

[68] Yu, S., Piao, X., Yoo, K., Shin, J. and Park, N. (2015). One-way optical modal transition based on causality in momentum space. *Opt. Express*, 23: 24997–25008.

[69] Longhi, S. (2016). Bidirectional invisibility in Kramers-Kronig optical media. *Opt. Lett*, 41: 2727–2730.

[70] Phang, S., Vukovic, A., Susanto, H., Benson, T. M. and Sewell, P. (2014). Ultrafast optical switching using PT-symmetric Bragg gratings. *J. Opt. Soc. Am. B*, 30: 2984–2991.

[71] Fan, S., Baets, R., Petrov, A., Yu, Z., Joannopoulos, J. D., Freude, W., Melloni, A., Popovic, M., Vanwolleghem, M., Jalas, D., Eich, M., Krause, M., Renner, H., Brinkmeyer, E. and Doerr, C. R. (2012).

Comment on 'Nonreciprocal light propagation in a silicon photonic circuit.' *Science*, 335: 38.

[72] Haus, H. A. (1984). *Waves and Fields in Optoelectronics* (Prentice-Hall).

[73] Bohren, C. F. and Huffman, D. R. (1998). *Absorption and Scattering of Light by Small Particles* (Wiley).

[74] Thiessen, E., Heinisch, R. L., Bronold, F. X. and Fehske, H. (2016). Surface mode hybridization in the optical response of core-shell particles. *Phys. Rev. A*, 93: 033827.

[75] Shen, Y., Wang, L. V. and Shen, J. (2014). Ultralong photonic nanojet formed by a two-layer dielectric microsphere. *Opt. Lett*, 39: 4120–4123.

[76] Mirzaei, A., Shadrivov, I. V., Miroshnichenko, A. E. and Kivshar, Y. S. (2013). Cloaking and enhanced scattering of core-shell plasmonic nanowires. *Opt. Express*, 21: 10454–10459.

[77] Razeghi, M. and Nguyen, B. (2014). Advances in mid-infrared detection and imaging: a key issues review. *Rep. Prog. Phys.*, 77: 082401.

[78] Soci, C., Zhang, A., Xiang, B., Dayeh, S. A., Aplin, D. P. R., Park, J., Bao, X. Y., Lo, Y. H. and Wang, D. (2007). ZnO nanowire UV photodetectors with high internal gain. *Nano Lett.*, 7: 1003–1009.

[79] Liu, X., Gu, L., Zhang, Q., Wu, J., Long, Y. and Fan, Z. (2014). All-printable band-edge modulated ZnO nanowire photodetectors with ultra-high detectivity. *Nat. Commun*, 5:4007.

[80] Li, R., Wei, Z., Zhao, F., Gao, X., Fang, X., Li, Y., Wang, X., Tang, J., Fang, D., Wang, H., Chen, R. and Wang, X. (2017). Investigation of localized and delocalized excitons in ZnO/ZnS core-shell heterostructured nanowires. *Nanophotonics*, 7: 1093–1100.

[81] Jackson, J. D. (1998). Classical Electrodynamics (Wiley).

[82] Alu, A., Rainwater, D. and Kerkhoff, A. (2010). Plasmonic cloaking of cylinders: finite length, oblique illumination and cross-polarization coupling. *New J. Phys.*, 12: 103028.

[83] Monticone, F., Argyropoulos, C. and Alu, A. (2013). Multilayered plasmonic covers for comblike scattering response and optical tagging. *Phys. Rev. Lett.*, 110: 113901.

[84] Monticone, F., Argyropoulos, C. and Alu, A. (2012). Layered plasmonic cloaks to tailor the optical scattering at the nanoscale. *Sci. Rep.*, 2: 912.

[85] Silveirinha, M. G., Alu, A. and Engheta, N. (2008). Infrared and optical invisibility cloak with plasmonic implants based on scattering cancellation. *Phys. Rev. B*, 78: 075107.

[86] Aryaeepanah, M. E., Takayama, O., Morozov, S. V., Kudryavtsev, K. E. Semenova, E. S. and Lavrinenko, A. V. (2016). Highly doped InP as a low loss plasmonic material for mid-IR region. *Opt. Express*, 24: 29077–29088.

[87] Farhat, M., Muhlig, S., Rockstuhl, C. and Lederer, F. (2012). Scattering cancellation of the magnetic dipole field from macroscopic spheres. *Opt. Express*, 20: 13896–13906.

[88] Kim, K., No, Y., Chang, S., Choi, J. and Park, H. (2015). Invisible hyperbolic metamaterial nanotube at visible frequency. *Sci. Rep.*, 5: 16027.

[89] Diaz-Avino, C., Naserpour, M. and Zapata-Rodriguez, C. J. (2016). Optimization of multilayered nanotubes for maximal scattering cancellation. *Opt. Express*, 24: 18184–18196.

[90] Huang, Y., Min, C., Dastmalchi, P. and Veronis, G. (2015). Slow-light enhanced subwavelength plasmonic wave guide refractive index sensors. *Opt. Express*, 11: 14922–14936.

[91] Zhang, L., Tu, R. and Dai, H. (2006). Parallel core-shell metal-dielectric-semiconductor germanium nanowires for high-current surround-gate field-effect transistors. *Nano Lett.*, 6: 2785–2789.

[92] Wiecha, P. R., Cuche, A., Arbouet, A., Girard, C., Francs, G., Lecestre, A., Larrieu, G., Fournel, F., Larrey, V., Baron, T. and Paillard, V. (2017). Strongly directional scattering from dielectric nanowires. *ACS Photonics*, 4: 2036–2046.

[93] Yariv, A., and Yeh, P. (2007). *Optical Electronics in Modern Communications* (Oxford University).

[94] Veronis, G., Dutton, R. W. and Fan, S. (2004). Method for sensitivity analysis of photonic crystal devices. *Opt. Lett.*, 29: 2288–2290.
[95] Jin, J. (2002). *The Finite Element Method in Electromagnetics* (Wiley).
[96] Taflove, A. (1995). *Computational Electrodynamics* (Artech House).
[97] Babicheva, V. E., Kulkova, I. V., Malureanu, R., Yvind, K. and Lavrinenko, A. V. (2012). Plasmonic modulator based on gain-assisted metal-semiconductor-metal waveguide. *Photon. Nanostructures*, 10: 389–399.
[98] Levinshtein, M., Rumyantsev, S. and Shur, M. (1999). *Handbook Series on Semiconductor Parameters* (World Scientific).
[99] Rafailov, E. U., Cataluna, M. A. and Sibbett, W. (2007). Mode-locked quantum-dot lasers. *Nat. Photonics*, 1: 395–401.
[100] Kawaguchi, K., Yasuoka, N., Ekawa, M., Ebe, H., Akiyama, T., Sugawara, M. and Arakawa, Y. (2008). Demonstration of transverse-magnetic dominant gain in quantum dot semiconductor optical amplifiers. *Appl. Phys. Lett.*, 93: 121908.
[101] Semenova, E. S., Kulkova, I. V., Kadkhodazadeh, S., Schubert, M. and Yvind, K. (2011). Metal organic vapor-phase epitaxy of InAs/InGaAsP quantum dots for laser applications at 1.5 μm. *Appl. Phys. Lett.*, 99: 101106.
[102] Kats, M. A., Sharma, D., Lin, J., Genevet, P., Blanchard, R., Yang, Z., Qazilbash, M. M., Basov, D. N., Ramanathan, S. and Capasso, F. (2012). Ultra-thin perfect absorber employing a tunable phase change material. *Appl. Phys. Lett.*, 101: 221101.
[103] Kats, M. A., Blanchard, R., Genevet, P. and Capasso, F. (2013). Nanometre optical coatings based on strong interference effects in highly absorbing media. *Nat. Mater.*, 12: 20–24.
[104] Nezhad, M. P., Tetz, K. and Fainman Y. (2004). Gain assisted propagation of surface plasmonpolaritons on planar metallic waveguides. *Opt. Express*, 12: 4072–4079.
[105] Bimberg, D., Ledentsov, N. N., Grundmann, M., Heinrichsdorff, F., Ustinov, V. M., Kopev, P. S., Alferov, Z. I. and Lott, J. A. (1998).

Edge and vertical cavity surface emitting InAs quantum dot lasers. *Solid St. Electron*, 42: 1433–1437.

[106] Kirstaedter, N., Schmidt, O. G., Ledentsov, N. N., Bimberg, D., Ustinov, V. M., Egorov, A. Yu., Zhukov, A. E., Maximov, M. V., Kopev, P. S. and Alferov, Z. I. (1996). Gain and differential gain of single layer InAs/GaAs quantum dot injection lasers. *Appl. Phys. Lett.*, 69: 1226–1228.

[107] Akahane, K., Ohtani, N., Okada, Y. and Kawabe, M. (2002). Fabrication of ultra-high density InAs-stacked quantum dots by strain-controlled growth on InP(3 1 1)B substrate. *J. Cryst. Growth*, 245: 31–36.

[108] Akahane, K., Yamamoto, N. and Kawanishi, T. (2011). Fabrication of ultra-high-density InAs quantum dots using the strain-compensation technique. *Phys. Status Solidi A*, 208: 425–428.

[109] Dembowski, C., Graf, H., Harney, H. L., Heine, A., Heiss, W. D., Rehfeld, H. and Richter, A. (2001). Experimental observation of the topological structure of exceptional points. *Phys. Rev. Lett.*, 86: 787–790.

[110] Zhen, B., Hsu, C. W., Igarashi, Y., Lu, L., Kaminer, I., Pick, A., Chua, S., Joannopoulos, J. D. and Soljacic, M. (2015). Spawning rings of exceptional points out of Dirac cones. *Nature*, 525: 354–358.

[111] Doppler, J., Mailybaev, A. A., Bohm, J., Kuhl, U., Girschik, A., Libisch, F., Milburn, T. J., Rabl, P., Moiseyev, N. and Rotter, S. (2016). Dynamically encircling an exceptional point for asymmetric mode switching. *Nature*, 537: 76–80.

[112] Liu, Q., Wang, B., Ke, S., Long, H., Wang, K. and Lu, P. (2017). Exceptional points in Fano-resonant grapheme metamaterials. *Opt. Express*, 25: 7203–7212.

In: Phase Change Materials
Editor: Ismaël van der Winden

ISBN: 978-1-53617-536-3
© 2020 Nova Science Publishers, Inc.

Chapter 3

A REVIEW ON THE THERMAL BEHAVIOR OF PHASE CHANGE MATERIALS INTEGRATED IN BUILDING ROOFS

I. Hernández-Pérez[*] *and J. Triano-Juárez*

División Académica de Ingeniería y Arquitectura, Universidad Juárez Autónoma de Tabasco, Cunduacán, Tabasco, México

ABSTRACT

The materials of the building envelope significantly influence heat gains and heat losses. The thermal protection offered by building materials consists of insulating, reflecting, or storing heat. Based on heat storage, phase change materials (PCM) use the heat gain for their phase change, which requires large amounts of energy per unit of volume. This feature allows increasing the heat storage capacity of the building envelope in the order of tens of times. For specific weather conditions, the thermal efficiency of a PCM when integrated in the building envelope depends on the phase change temperature, the latent heat, the volume of material, the position within the building envelope, and the orientation of the PCM, as well as the thermophysical properties of the other elements

---

[*] Corresponding Author's E-mail: ivan.hernandezp@ujat.mx.

of the building envelope. Because building roofs have prolonged exposure to solar radiation, they are one of the primary sources of energy gains, especially in regions with warm weather. Thus, this chapter aims to review the application of PCMs in building roofs. The design parameters considered when analyzing building roofs with PCM are mentioned as well as which parameters have shown the more significant influence on the thermal behavior of roofs. Several roof-PCM configurations reported in the literature are illustrated. Further, this chapter discusses how effective are those configurations to reduce energy consumption and to improve indoor comfort conditions in buildings.

Keywords: PCM, building roofs, energy consumption, thermal comfort

1. INTRODUCTION

The buildings sector is responsible for 36% of total final energy use around the world, more than half of global electricity consumption, and 39% of energy-related CO_2 emissions (UN 2017). Because, the population is expected to increase by 2.5 billion people in 2050, the energy use in the building sector is set to rise sharply (IEA 2013).

The building envelope is the boundary between the outdoors and the indoor environment of a building. Thus, the design and selection of the building materials that integrate the building envelope significantly influence thermal comfort conditions (Pacheco-Torgal et al. 2017). Furthermore, the trend of modern architecture aimed at the construction of non-massive and highly glazed buildings causes a significant consumption of electricity due to the high cooling loads, which also causes an increase of the greenhouse gas emissions (Lira-Oliver and Vilchis-Martínez 2017; Zhu et al. 2018; Karaoulis 2017). Given this panorama, it has been proposed as one of the axes of energy efficiency to improve the thermal design of buildings through the proper selection of construction materials, in such a way that the building envelope contributes to achieving the thermal comfort of the inhabitants with a minimum energy consumption (Pacheco-Torgal et al. 2017). The thermophysical and optical properties, as well as the thickness of each layer of the building components, influence the building energy gain. Thus, one can take advantage of the capacity that some materials have to reflect, delay, or store the incident energy to have an energy-efficient building.

One of the measures most commonly used today to reduce heat gain focuses on reducing the overall heat transfer coefficient of the building envelope by adding insulating materials. This method can significantly reduce heat transfer between outdoor and indoor environments (Yuan et al. 2017; Martin et al. 2018). Another effective measure to reduce heat gain is the application of reflective coatings on the outer surface of the envelope. With this method, it is possible to reduce the absorption of solar radiation by an opaque surface, however, the value of the solar reflectance of the coatings decreases over time due to factors such as soiling, pollution, weather type, among others (Paolini et al. 2014; Sleiman et al. 2011).

Conventional building materials such as concrete, brick, among others, used in building envelopes can store thermal energy. In these materials, stored heat causes an increase in temperature (sensible heat) that is proportional to the amount of energy stored and inversely proportional to the mass of the material (Fan and Luo 2018). In the last decade, several researchers have shown that phase change materials (PCM) technology can be used in the building envelope to store incident solar energy during the day and release it to the outdoor environment during the night (Aketouane et al. 2018). PCMs use the stored heat to perform their phase change, which occurs at a constant temperature range and requires a large amount of energy per unit volume. This feature allows the PCM to have a heat storage capacity tens of times higher than concrete without affecting the thickness of the envelope. As a result, it is possible to reduce heat gains, attenuate temperature fluctuations, retard the maximum outside surface temperature during a period known as time lag, and decrease the peak heat flux at a coefficient known as decrement factor. However, for optimal use of storage capacity, the PCM must be fully melted during daylight hours and solidified during nighttime hours, in other words, the PCM must complete its daily phase change cycle (Lira-Oliver and Vilchis-Martínez 2017; Zhu et al. 2018; Zwanzig, Lian, and Brehob 2013; Tokuç, Yesügey, and Başaran 2017; Dong et al. 2015; Reddy, Mudgal, and Mallick 2017).

For application in buildings, PCMs are available according to their phase change temperature, type of PCM (organic, inorganic, and eutectic), and type of packaging (micro or macro-encapsulated). Organic PCMs have qualities that make them suitable for buildings such as paraffin, whose use is the most frequent regardless of the climatic zone (maximum frequency of use of 87.5%) (Cui et al. 2017). Micro-encapsulated PCMs can expand their volume without affecting the integrity of the envelope and can be incorporated directly, for example, into concrete or plaster. The macro-encapsulation of the PCM in tubes, spheres, or panels is the most widely used technique in the building sector due to its extensive design and manufacturing flexibility (Kalnæs and Jelle 2015; Bland et al. 2017).

PCMs can be incorporated into any component of the building envelope (walls, roof, and floor), however, under warm weather conditions, roofs can significantly influence the energy consumption in buildings due to their direct and prolonged exposure to the sun. According to Nahar et al. 2003, a roof can contribute with up to 50% of the thermal load for buildings in warm locations. The efficiency of the PCM depends on several design parameters such as: the phase change temperature, latent heat, volume, position or distribution of the PCM (use of one or multiple layers), as well as the thermophysical properties of the others materials that make up the roof. Therefore, most of the researches avalailable in the literature have focused on finding the most favorable roof configuration for the climatic conditions of a particular region. The most frequently analyzed parameters when studying building roofs integrated with PCM are the phase change temperature, the thickness and the position of the PCM layer (when it is in the form of panels or plates). The influence of these parameters on the thermal performance of building roofs is described in this work. Therefore, the chapter presents the state of the art of the research with regard to the thermal performance of PCMs integrated in building roofs. Different methodologies have been used to study the potential of PCMs to reduce energy consumption and to improve the thermal comfort. Therefore, the chapter is divided into two major sections, studies in air-conditioned buildings and studies on non air conditioned buildings. In each section, a summary table describes the roof configurations, the location of the study, the characteristics of the models, and the main results.

2. THERMAL BEHAVIOR OF PCM INCORPORATED IN ROOFS OF AIR CONDITIONED BUILDINGS

In warm locations, the air temperature reached inside buildings usually exceeds the values considered within the thermal comfort zone. Thus, most of the buildings in these locations have an air conditioner to provide thermal comfort to their inhabitants. Extensive research related to the implementation of PCM into building roofs has been carried out in air-conditioned buildings to reduce cooling energy loads. Some researchers have directly determined this effect. Others have studied the behavior of different variables such as the interior surface temperature, the heat flux, or the heat gain of the roofs, which influence the cooling energy loads. Therefore, when a roof with PCM reduces any of the variables mentioned above, it naturally reduces the energy consumption of a building.

2.1. Influence of PCM's in the Minimization of Cooling Energy Loads

The reduction of the cooling loads of air-conditioned buildings due to the incorporation of PCM in roofs was analyzed in two research works. In the first work, Tokuç et al., 2017 evaluated a roof with and without PCM in terms of the cooling load for the summer season in two weather zones of Turkey. The authors showed that, compared to a building without PCM, the roof with PCM with thicknesses of 1, 2, 3, 4, and 5 cm reduced the cooling loads in Izmir by 3.4, 6.4, 9, 11.4, and 13.6%, respectively. In Erzurum, the roof with the same PCM thicknesses reduced the cooling load by 20.1, 41.5, 49.2, 54.9, and 58.3%, respectively. Therefore, they concluded that the reductions of the cooling loads are more significant for larger thicknesses of the PCM layer. In the second work, Zwanzig et al., 2013 evaluated the effect of the location of a PCM layer inside a roof on the energy-saving potential of a residential building located in three climatic zones in the USA (Minneapolis, Louisville, and Miami). Three roof configurations were studied, 1) PCM layer near the exterior surface; 2) PCM layer next to the interior surface; 3) PCM layer on the interior surface. For the city of Minneapolis, the configurations 1, 2, and 3 reduced the cooling load up to 5.3, 9.1, and 11.4%, respectively. For the city of Louisville, the configurations 1, 2, 3 reduced the cooling load up to 3.7, 5.4, and 8.0%, respectively. Finally, for the city of Miami, the configurations 1, 2, 3 reduced the cooling load up to 2.7, 2.8, and 4.1%, respectively. More details of the two works presented above are shown in Table 1.

2.2. Influence of the PCM's Physical Features on the Thermal Behavior of Building Roofs

Several parametric studies were carried out to evaluate either the thickness or the location of a PCM layer on the thermal behavior of building roofs. Some of these studies consider that the PCM layer has different values of its thermophysical properties such as the phase change temperature, the phase transition temperature range, and the heat of fusion or the latent heat. The results from these parametric studies show that incorporating a PCM layer into building roofs reduces the internal surface temperatures, the heat flux, and the heat gain of roofs.

One of the first parametric evaluations of PCM incorporated into building roofs was published in Dong et al., 2015. The authors studied the thermal performance of a roof with and without PCM situated in the northeast cold area of China. Compared to the roof without PCM, the roof with a PCM layer of 10 cm thick and latent heat of 188 kJ·kg^{-1} reduced the temperature of the interior surface by 10, 13, and 12°C and the net heat gain by 7.36, 7.04, and 7.52 W·m^{-2}, when the PCM had a phase change temperature of 303 K, 307 K, and 311 K, respectively. Moreover, the authors identified that the different melting temperatures lead the PCM to complete its phase change in a particular way. They found that the average percentage at which each PCM completed its phase change was 18.1, 28.9, and 26%, for the PCM with a melting temperature of 303, 307, and 311 K, respectively. For each of the phase change temperatures analyzed, three heat of fusion (138, 188, and 238 kJ·kg^{-1}) were tested maintaining a constant thickness of 10 cm in the PCM layer. The temperature distribution of the internal surface temperature of the different latent heat of each phase change temperature showed maximum differences of 0.5°C regarding the maximum reductions mentioned before. However, in the average percentage in which each PCM completed its phase change maximum differences of 10, 14.2, and 11.2% were obtained for PCMs with phase change temperatures of 303 K, 307 K and 311 K, respectively. For the case with PCM with a melting temperature of 311 K, a heat of fusion equal to 188 kJ·kg-1, and a thickness of 10 cm, the authors tested different slopes (0.05, 0.15, 0.33, 0.4, and 0.5). The relative difference between the maximum values of the interior surface temperature and the heat flux, as well as the average percentage of the liquid fraction between the minimum and maximum values of the slope (0.05-0.5) were 2.5°C, 7 W·m^{-2} y 29%, respectively. Therefore, compared to the case without PCM, the variation of the slopes reduced the maximum temperature of the interior surface between 13-13.8°C. In addition, for the thicknesses of 4, 6, 8, and 10 cm of the same PCM layer, reductions were obtained in the maximum temperature of the interior surface of 9.5, 12, 12.5, and 13°C, respectively; and average percentages of liquid fraction of 79, 52, 36, and 26%. With the same PCM the maximum temperature of the internal surface was reduced 12, 12, 10 and 4°C for values of solar absorptivity of the external surface of the roof of 0.6, 0.7, 0.8 and 0.9, respectively.

A second parametric study carried out in the humid tropical weather of Chennai, India was developed by Reddy et al., 2017. They simulated the thermal behavior of a roof integrated with a PCM. The indoor air temperature of the simulated building remained constant at 27°C. The researchers evaluated three PCMs with different melting and heat of fusion, four thicknesses of the PCM layer, and the application of multiple layers of PCM. In all cases of study, the PCM was located near to the internal surface of the roof.

Table 1. Characteristics of the studies that have determined the influence of PCM's on the cooling energy loads

Reference	Weather conditions	Roof configuration	T_{pc} (°C)	Thickness of the roof (cm)	Main results
Tokuç et al., 2017	Climatic zones of Turkey	Ceramic and mortar 1.5 cm; Concrete 5 cm; Water insulation 1 cm; PCM 5 cm; Extruded polystyrene 6 cm; Reinforced concrete 10 cm	25.72-26.22	28.5	During the period May-September the largest decrease in the cooling load, 58%, was obtained for the city of Erzurum, followed by Ankara with 24.3%, Istanbul with 15.3%, and Izmir with 13.6%.
Zwanzig et al., 2016	Weather of Louisville, Miami, and Mineapolis	Built-up roofing 0.95 cm; Fiberboard sheathing 1.27 cm; Insulation board 5.08 cm; Wood 5.08 cm; PCM 1.27 cm	25-27.5	13.65	For several days of July, the peak het flux was reduced by 42% in Louisville, by 33.5% in Miami, and by 33% in Mineapolis with reductions of 8, 4, and 11% of the cooling load, respectively.

T_{pc}: Phase change temperature

Compared to the base case (roof without PCM), the roof case with PCM with a melting temperature of 300-302 K and a latent heat of 132 kJ·kg^{-1} reduced the maximum temperature of the internal surface of the roof by 4°C. The case with PCM with a melting temperature of 303.4-305 K and a latent heat of 188 kJ·kg^{-1} reduced the maximum internal surface of the roof by 6.6°C. Further, the case with PCM with a melting temperature of 306-307 K and a latent heat of 197 kJ·kg^{-1} reduced the maximum temperature of the internal surface of the roof by 4.3°C. The analysis of the effect of the thicknesses was performed with the PCM with a melting temperature of 300-302 K, where the reductions of maximum surface temperatures of 4, 4.5, 5.7, and 6.7°C were obtained for the thicknesses of 2, 4, 6, and 8 cm, respectively. Regarding the effect of the number of PCM layers, they showed that a roof with a 2 cm thick layer of PCM with a melting temperature of 300-302 K reduced the heat gain between 17 and 26%, and by adding another 3 cm thick layer of PCM with melting temperature of 303.4-305 K (double PCM layered roof) the heat gain was reduced between 25 and 36%.

Another study of the same group was developed by Akeiber et al., 2017, who studied the influence of PCM panels on the heat transfer of the building envelope. The researchers built two identical rooms, one used as the reference, and the other used to test the PCM technology installed as a layer in the walls and the roof. They tested three different thicknesses, 2.5, 4, and 6 cm of the PCM layer with a melting temperature range of 40-44°C. The indoor air temperature of the rooms was kept constant at 24°C by using an air conditioner. The room with PCM decreased heat flux 74.8, 38.5, and 59.3% for thicknesses of 6, 4, and 2.5 cm, respectively.

Research carried out by Mourid and El Alami, 2017, has experimentally studied the thermal performance of a PCM incorporated in a roof under the weather conditions of Casablanca, Morocco. The authors studied the effect of the thickness and location of the PCM layer, with a melting temperature range of 21.7-31°C, on the heat transfer of the roof. When analyzing the effect of the thickness, they found that the roof configuration with one layer of PCM (5.26 mm) compared to a roof without PCM reduced the peak heat flux by 58%, while the configuration with two layers of PCM reduced the peak heat flux by 88%. In both configurations, the PCM layers were placed near the outer surface of the roof. When analyzing the effect of the position, they found that the roof configuration with the PCM adjacent to the inner surface of the roof was more effective. This configuration decreased both the heat flux and the inner surface temperature of the roof by 49% and 3°C, respectively, when compared to the roof configuration with the PCM adjacent to the outer surface of the roof.

Another parametric study has investigated the effect of three physical features of the PCM layer (phase change temperature, thickness, the phase transition temperature range) on the peak interior surface temperature of roofs (Yu et al., 2019). The researchers studied a roof model with a layer of PCM incorporated on the external surface under the weather conditions of Wuhan, China. When analyzing the phase change temperature, they considered that the layer of PCM had a thickness of 3 cm. The results of the different cases tested were compared with a reference case (a roof that replaces the PCM layer with a 3 cm thick mortar). The roofs with a 3 cm thick PCM layer with the phase change temperature ranges of 31-33°C, 33-35°C, 34-36°C, 39-41°C, 35-37°C, 37-39°C, and 36-38°C, reduced the internal peak surface temperature by 2, 2.6, 3, 3.2, 3.5, 3.7, and 3.9°C, respectively. When analyzing the thickness of the PCM, they used a PCM layer with the best phase change temperature range (36-38°C). The results showed that the tested thicknesses of 2, 3, 4, and 5 cm caused peak interior surface temperature reductions of 2.7, 3.9, 4.1, and 4.2°C, respectively. When analyzing the phase transition temperature range (ΔT = 4 and 6°C), they also considered that the layer of PCM had a thickness of 3 cm. The results showed that with the phase transition temperature range of 4°C, considered within the phase change temperature ranges of 34-38°C, 35-39°C, 36-40°C, 37-41°C, the peak interior surface temperatures were reduced up to 3.5, 3.7, 3.7, and 3.5°C, respectively. Further, with the phase transition temperature range of 6°C, which was considered within the phase change temperature ranges of 33-39, 34-40, 35-41, and 37-43°C reduced the peak interior surface temperatures up to 3.4, 3.6, 3.6, 3.3, respectively.

A novel roof configuration that combines PCM and ventilation was recently proposed by Yu et al., 2020. The authors analyzed the thermal insulation performance of an innovative pipe-embedded ventilation roof with outer-layer shape-stabilized PCM. The study was conducted for the conditions of a typical summer day in five representative climatic regions of China. The first parameter evaluated was the airflow rate in the ventilation duct, using a 3 cm thick PCM layer. The results obtained in Harbin (severe cold zone) with a PCM with phase change temperature of 31-33°C, showed that compared to the case without ventilation, cases with ventilation speeds of 0.4, 0.8, 1.2 and 2.0 ms^{-1} reduced the minimum internal surface temperature by 1.28, 2.2, 2.72, 3.36°C; and the maximum temperature of the internal surface by 0.56, 0.64, 0.70 and 0.77°C, respectively. For a PCM with a phase change temperature of 34-36°C at Beijing conditions (cold zone), cases with ventilation speeds of 0.4, 0.8, 1.2 and 2.0 ms^{-1} compared to the case without ventilation, reduced the temperature minimum internal surface 0.85, 1.4, 1.75 and 1.92°C. In these cases, it was observed that the ventilation rate has little effect on the maximum temperatures of the internal surface (the values are approximately equal). In Wuhan (hot-summer and cold-winter zone) using a PCM with phase change temperature of 36-38°C and air ventilation speeds of 0.4, 0.8, 1.2 and 2.0 ms^{-1}, the minimum temperature of the internal surface, compared to the case without ventilation, was reduced 0, 0.3, 0.45 and 0.55°C. The results showed a slight effect in lowering the maximum internal surface temperature. The results obtained in Guangzhou (hot-summer and warm-winter zone) with a PCM with a phase change temperature of 34-36°C, showed that, compared to the case without ventilation, cases with ventilation speeds of 0.4, 0.8, 1.2 and 2.0 ms^{-1} reduced the minimum internal surface temperature 0.4, 0.85, 1 and 1.3°C. Maximum internal surface temperatures showed reductions between 0.25-0.3°C. In Kunming (mild zone) with a PCM with phase change temperature of 29-31°C, compared to the case without ventilation, cases with ventilation speeds of 0.4, 0.8, 1.2 and 2.0 ms^{-1} reduced the minimum temperature of the internal surface 2.3, 3.4, 4, and 4.8°C; and the maximum temperature of the interior surface 1, 1.25, 1.45, and 1.5°C, respectively. The second parameter evaluated was the thickness of the PCM layer, which was determined by the latent heat utilization rates of the PCM (the difference between the highest and the lowest liquid fraction of the PCM during a day). For the evaluation, ventilation speeds of 1.6 m/s, 2.0 m/s, 2.4 m/s, 2.0 m/s, and 1.6 m/s were used in Harbin, Beijing, Wuhan, Guangzhou, and Kunming respectively. For a thickness of 2 cm, the latent heat utilization rate was 100, 100, 92, 100, and 100% in Harbin, Beijing, Wuhan, Guangzhou, and Kunming, respectively. For a thickness of 2.5 cm, the latent heat utilization rate was 79, 88, 100, 83, and 58%, respectively. And for a thickness of 3 cm, the latent heat utilization rate was 51, 54, 100, 51, and 33%, respectively. Table 2 summarizes the characteristics of the studies presented in this section.

2.3. Influence of the PCM on the Time Lag and Decrement Factor of Building Roofs

Some recent works have studied the effect of a PCM layer on the time lag (TL) and the decrement factor (DF) of building roofs. Some of these researches also studied the influence of PCMs in the internal surface temperatures, the heat flux, and the heat gain of roofs. A study described in Kharborouc et al., 2018 has experimentally and numerically studied the thermal performance of a PCM integrated into a roof exposed to weather conditions of Tangier, Morocco. In the numerical part, a fixed indoor air temperature (23°C) was established. The effect of the phase change temperature and the location of the PCM layer was evaluated by using the time lag and the decrement factor. For a 1 cm thick PCM layer located near the inner surface and with melting temperatures of 21, 23, 25, and 27°C, the roof had DF of 0.049, 0.031, 0.034, and 0.050, and TL of 308, 466, 462, 345 min, respectively. When analyzing the effect of the location, the researchers found that the roof with a PCM layer located 1.3 cm away from the interior surface had a DF of 0.031 and a TL of 465 min. Further, they showed that when the PCM was located 19 cm away from the exterior surface, the DF and TL were 0.040 and 315 min, respectively.

A second study carried out for the conditions of five regions with different weather in China (Harbin, Beijing, Guangzhou, and Kunming) is described in Yu et al., 2019. The authors simulated the thermal behavior of an outer-layer shape-stabilized phase change roof to determine the optimal phase change temperature. The simulated period was a typical summer day. In all cases, the 3 cm thick PCM layer was located 2 cm from the outer surface.

Table 2. Characteristics of the studies that have determined influence of the PCM's physical features on the thermal behavior of building roofs

Reference	Weather conditions	Roof configuration	T_{pc} (°C)	Thickness of the roof (cm)	Main results
Dong et al., 2015	Northeast and cold zone of China $T_{\infty,máx} \approx 29°C$, $I_{sol\,máx} \approx 600$ Wm^{-2}	Aluminum alloy / Concrete / PCM / Reinforced concrete — 10 cm, 10 cm, 2 cm, 0.2 cm	34	22.2	For a June day, the PCM caused a reduction of 13°C the of the roof internal surface, a time lag of 3 h, and a reduction of 11% of the heat gain. On average, the PCM completed its phase change process by around 29%.
Reddy et al., 2017	Weather of Chennai, India $T_{\infty,max} \approx 35°C$, $I_{sol\,max} \approx 1000$ Wm^{-2}	Brick 8 cm / Concrete 8 cm / PCM 305K 3 cm / PCM 302K 2 cm / Gypsum board 2 cm	30-32 27-29	23	The double layer of PCM achieved its complete melt-solidification cycle, which resulted in a constant temperature of 28°C on the internal surface of the roof during a day in July. The one-year simulation showed a reduction of the heat gain between 25 and 36%.
Akeiber et al., 2017	Weather of Iraq (a whole day with outdoor temperature between 36-44°C)	Mortar + brick mixture 10 cm / PCM 6 cm / Concrete 12 cm	40-44	28.0	From 0:00 a.m. to 1:00 p.m., the temperature of the lower surface of the roof remained at 29°C and had a peak of 32°C. The heat flux was reduced by 74.75% due to the PCM.

Table 2. (Continued)

Reference	Weather conditions	Roof configuration	T_{pc} (°C)	Thickness of the roof (cm)	Main results
Mourid and El Alami, 2017	Weather of Casablanca, Morocco $T_{\infty,max} \approx 37°C$, $I_{sol,max} \approx 900$ W m^{-2}	PCM 0.526 cm / PCM 0.526 cm / Cement mortar 2 cm / Bitumen 3 cm / Cement mortar 2 cm / Heavy Concrete 15 cm / Cement mortar 2 cm	21.7-31	25.052	For three days of summer the incorporation of two layers of PCM of equal thickness and T_{pc}, produced a reduction of the flux of heat of 88%, and the temperature of the interior surface of the ceiling ranged between 24.5 and 28°C.
Yu et al., 2019	A tipical day of summer of five climatic zones of China (Harbin, Beijing, Wuhan, Guangzhou, and Kunming).	Cement mortar 2 cm / PCM 3 cm / Cement mortar 2 cm / Concrete / Air Air 12 cm	19	Harbin: 31-33°C Beijing: 34-36°C Wuhan: 36-38°C Guangzhou 34-36°C Kunming: 29-31°C	In severe cold region (Harbin), cold region (Beijing), hot summer cold winter region (Wuhan), hot summer warm winter region (Guangzhou) and mild region (Kunming), compared with roof without PCM, the decrement factors are decreased by 85.90%, 87.12%, 85.78%, 87.83% and 88.79%, the peak inner surface temperatures are decreased by 3.7, 4.0, 3.9, 3.8, and 3.7°C, and the inner surface temperature amplitudes are decreased by 6.2, 6.5, 6.1, 6.2, and 6.7°C, respectively.

Reference	Weather conditions	Roof configuration	T_{pc} (°C)	Thickness of the roof (cm)	Main results
Yu et al., 2020	A tipical day of summer of five climatic zones of China (Harbin, Beijing, Wuhan, Guangzhou, and Kunming).	Cement mortar 2 cm / PCM 3 cm / Cement mortar 2 cm / Concrete / Air Air 12 cm	19	Harbin: 31-33°C Beijing: 34-36°C Wuhan: 36-38°C Guangzhou: 34-36°C Kunming: 29-31°C	In Harbin, Beijing, Wuhan, Guangzhou, and Kunming the suitable airflow rates were 1.5-1.9, 1.6-2.0, 2.1-2.5, 1.9-2.3, and 1.4-1.8 ms^{-1}, respectively. Compared to the non-ventilated condition, the heat gain through the roof in summer typical day was reduced by 1604.45, 949.54, 300.67, 565.06, and 2410.56 kJm^{-2}, respectively.

T_{pc}: Phase change temperature; $T_{\infty,max}$: Maximum ambient temperature; $T_{\infty,min}$: Minimum ambient temperature; $I_{sol,max}$: Maximum solar irradiance.

For Wuhan it was obtained that, compared to the reference case (roof without PCM), the use of PCMs with phase change temperature ranges of 31-33, 33-35, 34-36, 35-37, and 37-39°C reduced the maximum temperature of the inner surface of the roof 2, 2.6, 3.5, 3.9, and 3.7°C, respectively. The DF decreased 40.5, 55.6, 79.3, 85.7, and 75.4%, and the TL increased 2, 2, 2, 3, and 3 h, respectively. For Haerbin, compared to the reference case, the use of PCMs with phase change temperature ranges of 30-32, 31-33, 32-34, 33-35, and 35-37°C, reduced the maximum temperature of the inner surface of the roof 3, 3.6, 3.7, 3.47 and 2.9°C, respectively. The DF decreased 72.2, 85.9, 83.3, 71.8, and 55.5%, respectively, and the TL increased 1, 1, 1, 2, and 2 h, respectively.

Similarly, in Beijing, reductions of 2.53, 3.44, 3.94, 3.8, 3.27°C were obtained in the maximum temperature of the interior surface of the roof by incorporating PCMs with ranges of the temperature of phase change of 31-33, 33-35, 34-36, 35-37, and 37-39°C, respectively. In these cases, the DF decreased 55.4, 76.8, 87.1, 77.7, and 60.0%, respectively, and the TL increased 1, 1, 1, 2, and 2 h, respectively. Compared to the results of the

Guangzhou reference case, cases with PCM with phase change temperature ranges of 31-33, 33-35, 34-36, 35-37 and 37-39°C, reduced the maximum temperature of the inner surface of the roof 2.36, 3.26, 3.75, 3.63, 3.11°C, and DF 54, 77, 87.8, 79, 51.73%, respectively. In these cases, the TL increased 2, 2, 2, 3, and 3 h, respectively. For the Kunming region, reductions of 2.35, 3.26, 3.7, 3.5, and 3°C were obtained, when the PCM phase change temperature range was 26-28, 28-30, 29-31, 30-32, 32-34°C, respectively. In these cases, the DF decreased 57.7, 79.7, 88.8, 77.6, and 59.5% and the TL increased 2, 2, 3, 3, and 4 h, respectively.

In the third study of the same group, Triano-Juárez 2019 numerically analyzed the thermal behavior of a concrete roof with an internal layer of PCM. The author evaluated the effect of the phase change temperature and the position of a PCM layer on the thermal behavior of a roof situated in a city with warm-humid weather in Mexico. They compared the cases with PCM with the reference case (roof without PCM). The three positions evaluated correspond to the cases where the PCM layer, with a thickness of 1 cm, is close to the external surface (2 cm), in the middle, or close to the internal surface (2 cm). The results obtained for a PCM with a phase change temperature of 29-36°C showed that, in each of the previous cases, the corresponding average maximum internal surface temperature was reduced 4.1, 3.7, and 3.7°C; the peak flux heat 19.0, 17.3 and 17.1%; and the heat gain was reduced 13.9, 13.7, and 13.7%. Likewise, the TL increased 66, 35, and 26 min, respectively, and in the three cases, the PCM completed daily its phase change cycle. When the authors tested the other PCM with a phase change temperature range, 41-44°C, the results for the three positions of the PCM layer showed reductions of 5.2, 5.6, and 6°C in the average maximum temperature of the internal surface; of 24.3, 25.9, and 27.8% in peak heat flux; and of 13.6, 13.2 and 12.8% the heat gain, respectively. Further, the TL increased 55, 72, and 104 min; and the phase change cycles were completed in 100, 97, and 79.6% in each corresponding case. Table 3 summarizes the characteristics of the studies presented in this section.

Table 3. Characteristics of the studies that have determined influence of the PCM on the time lag and decrement factor

Reference	Weather conditions	Roof configuration	T_{pc} (°C)	Thickness of the roof (cm)	Main results
Kharbouch et al., 2018	Tánger, Marruecos $T_{\infty,max} \approx 27°C$, $I_{sol,max} \approx 1000$ Wm^{-2}	Concrete 4.0 cm; Roof slab of concrete 15.0 cm; Insulation cork 4.5 cm; PCM 1.0 cm; Gypsum plaster 1.3 cm	23	41-42	For a period of three days of summer, a time lag of 466 min and a decrement factor of 0.031 were obtained because of the PCM
Triano-Juárez et al., 2019	A typical summer week of Cunduacán, Mexico $T_{\infty,max} \approx 40\text{-}42°C$, $I_{sol,max} \approx 922\text{-}1106$ Wm^{-2}	Reinforced concrete 2 cm; PCM 1 cm; Reinforced concrete 10 cm; 1 cm / 1 cm	29-36	13	The PCM reduced the average maximum temperature of the internal surface by 4.1°C; the peak of flux heat by 19%; and the heat gain was reduced by 13.9%. The time lag increased 66 min, and in each day the PCM melted and solidified completely.

T_{pc}: Phase change temperature; $T_{\infty,max}$: Maximum ambient temperature; $T_{\infty,min}$: Minimum ambient temperature; $I_{sol,max}$: Maximum solar irradiance.

2.2. Thermal Behavior of Building Roofs with PCM and Reflective Coatings

Some authors have analyzed the thermal performance of roofs with PCM in combination with reflective coatings. Reflective or cool coatings are applied on the exterior surface of the roofs. Because of their high solar reflectance, this type of coatings reflect most of the solar radiation incident on the roofs. Three studies have analyzed the benefits of installing both PCM and reflective surfaces on building roofs. In the research developed by Lu et al., 2016 it was experimentally studied the effect of the application of a PCM and a reflective coating on the heat transfer of buildings located in Tianjin, China. The authors used three test rooms with different roof configurations each. The roof configuration 1 corresponded to a roof with PCM, the roof configuration 2 to a roof with a reflective coating, and the configuration 3 to a roof with PCM and reflective coating. The maximum temperatures of the interior surface for the configurations analyzed were similar; configurations 1, 2, and 3 had peak interior surface temperatures of 28.7, 29, and 28.4°C, respectively. The same occurred regarding the heat flux, the differences between the peaks of cases 1 and 3 and between the peaks of cases 2 and 3 were only 1.28 $W \cdot m^{-2}$ and 2.85 $W \cdot m^{-2}$, respectively. However, the time lag of the roof configurations was different; the average differences in the time lag between cases 1 and 3, and between cases 2 and 3 were 24 and 122 min, respectively.

The second study is the one presented by Piselli et al., 2019, who studied the thermal behavior of roofs with different PCMs and waterproofing membranes. The authors used two waterproofing membranes (one based on polyurethane and the other based on bitumen) and three PCMs with different phase change temperatures (25, 31, and 44°C). Experimental tests were performed under the weather conditions of Rome and Abu Dhabi. They showed that the roof configurations with PCM + polyurethane-based membrane were more effective in reducing the peak interior surface temperatures than the configurations with PCM + bitumen membrane. In Rome, the temperature reductions provided by the best configuration were 19, 16, and 10°C when the PCM had phase change temperatures of 25, 31, and 44°C, respectively. While the corresponding reductions provided by the same phase change temperatures in Abu Dhabi were 28, 22, and 10°C.

The third study available was performed by Saafi and Daouas, 2019. They numerically evaluated the feasibility of PCMs on roofs exposed to Tunisian weather conditions. For a PCM with a phase change temperature of 24°C, they studied the simultaneous effect of the PCM thickness and the solar reflectance of the roof. When the solar reflectance remained as 0.1, the results showed that, with respect to a case without PCM, a thickness of 3, 2, and 1 cm of the PCM layer reduced the annual cooling energy load by 14.4, 16, and 8%, respectively. When the solar reflectance remained as 0.9, all the analyzed cases, including the case without PCM, had minimal annual cooling energy load with values similar to each other. These results indicated that the effect of solar reflectance on cooling loads is more pronounced than the impact of the thickness of the PCM layer. Moreover, the reduction of the cooling loads when using a PCM layer in the roof is more critical for low values of solar reflectance. The authors also analyzed the effect of the melting temperature on the external surface temperature of a roof with a layer of 2 cm thick of PCM and a solar reflectance and emittance of 0.89. Compared to the reference case, the case with PCM with a melting temperature of 24, 26, and 28°C reduced the daily temperature fluctuation (the difference between the maximum and minimum temperature of the outer surface) 1.9, 3.2, and 5.3°C. According to these results, the authors observed that the incorporation of a PCM contributes to reducing the thermal stress on the cool roof surface due to temperature fluctuations. Table 4 summarizes the characteristics of the studies presented in this section

3. THERMAL BEHAVIOR OF PCM INCORPORATED IN ROOFS OF NON AIR CONDITIONED BUILDINGS

The research related to PCM incorporated in roofs of non-air conditioned buildings or buildings in free-floating conditions, is not as extensive as the research developed for air-conditioned buildings. However, significant results have been shown by several researchers regarding the benefits provided by the utilization of PCM.

Because the inside air temperature is the most direct representation of thermal comfort conditions, this variable is used to evaluate the thermal comfort conditions of non-air-conditioned building. Two studies are currently available focusing in the potential decrease of the indoor air temperature because of the incorporation of PCM in building roofs. Zhu et al., 2018 determined the effect of the location and the thickness of a layer of PCM on the thermal behavior of a building in Tianjin, China.

Table 4. Characteristics of the studies that have analyzed roofs with PCM and reflective coatings

Reference	Weather conditions	Roof configuration	T_{pc} (°C)	Thickness of the roof (cm)	Main results
Lu et al., 2016	Tianjin, China $T_{\infty,max} \approx 35°C$, $I_{sol,max} \approx 950$ Wm^{-2}	Reflective coating layer 0.3 cm; Crank-resistance mortar 2 cm; PCM 0.16 cm; Adhesive mortar 1 cm; Water insulation 1 cm; Glass-fiber cement polystyrene core 10.2 cm	29.5-31.8	14.66	The PCM, monitored during 4 days of August, maintained the temperature of the lower surface of the roof in the range of 27-28.5°C, a maximum time lag of 270 min, and a reduction of the heat gain of $\approx$ 15%.
Piselli, Castaldo, and Pisello 2019	Weather data of Rome and Abu Dabi	Polyurethane cool membrane 0.4 cm; PCM 1.5 cm; 12 m; Mineral wool insulation 5 cm; Wood slab 12 cm	22-26	30.9	For the interval from 12:00 pm to 3:00 pm on a representative summer day, the maximum temperature reduction between the external surface and the lower interface of the PCM was 19°C in Rome and 28°C in Abu Dhabi (compared with the same configuration, but without PCM).
Saafi and Daouas 2019)	A summer representative day of Tunisia Mediterranean weather	White elastomeric coating - cm; Sealing 0.5 cm; Cement plaster 2.5 cm; PCM 2 cm; Concrete 10 cm; Reinforced concrete 15 cm; Cement-lime plaster 2 cm	28	32	The temperature fluctuation (difference between the maximum and minimum temperature) of the roof outside surface decreases from 16.99°C without PCM to 11.64°C reducing the thermal stress on the cool roof surface due to temperature fluctuations.

T_{pc}: Phase change temperature; $T_{\infty,max}$: Maximum ambient temperature; $T_{\infty,min}$: Minimum ambient temperature; $I_{sol,max}$: Maximum solar irradiance.

The authors used 5 mm layer of PCM to find its best location inside the building envelope. When the PCM was located near to the interior surface, it decreased the indoor air temperature up to 5°C compared to a building without PCM. When considered in the middle of the envelope, the PCM decreased the indoor air temperature up to 5.5°C. On the contrary, when located near the exterior surface, the PCM did not reduce the indoor air temperature. On the other hand, the authors found that the different thickness of the PCM (2.5, 5, and 7.5 mm) had the same influence on the indoor air temperature. They reported that the three cases with different thickness had the same peak indoor air temperature (37.5°C). The second study available in the literature was developed by Hasan et al., 2018, who performed an experimental study to evaluate the influence of installing a 1 cm layer of PCM, with a melting temperature of 44°C, in the roof of a building room. The researchers performed a series of experiments under the weather conditions of Kut, Iraq, and used a building room without PCM as the reference. They found that the room with PCM had an indoor air temperature up to 2°C lower than the reference roof. Further, they determined that the energy gain of the roof with PCM was 7% smaller than the one of the roof without PCM.

Other studies do not have determined the influence of the PCM on the indoor air temperature of buildings, but they have studied other variables such as the interior surface temperature or the heat flux of the roofs which have an indirect effect on the indoor comfort conditions. Therefore, when a roof with PCM minimizes either the internal surface temperature or the heat flux, it indeed benefits the indoor thermal conditions. For instance, Elarga et al., 2017 analyzed a roof divided into three parts; one of reference (without PCM) and the other two parts with a layer of PCM with a melting temperature of 28°C and 35°C, respectively. Both sections of PCM consisted of a 1 cm thick panel. The authors evaluated the thermal performance of the roof during the summer season of Torino, Italy. The results showed that compared to the reference section, the sections with PCM with a melting temperature of 35 and 28°C reduced the maximum internal surface of the roof by 2 and 8°C.

The case with the PCM with phase change temperature of 28°C, according to the temperature profiles of the PCM layer, had the best use of the storage capacity of the PCM because during the daylight hours the PCM remained in phase transition, while the PCM with melting temperature of 35°C was mostly kept in solid-state. They also showed that roof configurations with PCM reduced the peak heat flux between 13 and 59%, which can provide acceptable indoor thermal conditions in non-air-conditioned buildings. Another available study is the one developed by Pasupathy et al., 2008, who analyzed the thermal performance of a PCM incorporated into the roof of a residential building located in Chennai, India. The authors studied the behavior of the roof for a whole year. They observed positive effects for the period from December to April. In this period, the interior surface of the roof with a layer of PCM with a thickness of 3 cm reached a maximum temperature of around 27°C. When the thickness of the PCM was 1 cm, the roof reached a maximum interior surface temperature of 34°C. However, for the period between May and November, the maximum temperatures of the interior surface of the roof with PCM were higher than the temperatures of a roof without PCM (at least 4°C higher). The study of Pasupathy et al., 2008 highlights the importance of developing annual exploratory simulation in design stages to guarantee that the incorporation of a PCM brings benefits to a building situated in a specific location. Finally, Yoon et al., 2018a studied the thermal performance of a roof with PCM under the summer of Seoul, South Korea. The PCM was incorporated into the roofs in a hollow layer of wood plastic composite. In the first stage, the authors analyzed PCM with different melting temperatures but the rooftops with the same solar reflectance on the external surface. Compared to a roof without PCM, they found that the roof with PCM reduced the peak interior surface temperature by 1.1 and 2.6°C when the PCM had melting temperatures of 26 and 44°C, respectively. In the second stage, the authors analyzed the same roof configurations but with different solar reflectance on the external surface.

The authors found that the roof with a brown coating and PCM, with a melting temperature of 44°C, reduced the maximum interior surface temperature by 9.7°C compared with the roof without PCM and without coating. The roof with a reflective white coating reduced the maximum internal surface temperature by 15.1 and 14.7°C when the melting temperature of the PCM was 44 and 26°C, respectively. Table 5 summarizes the characteristics of the studies presented in this section.

CONCLUSION

As observed in the works reported in this chapter, the phase change temperature, the location, and the thickness of the PCM layer are the main parameters studied in the thermal behavior of the PCMs applied in the roofs of the buildings. The objective of testing different configurations has been to identify the one that contributes to reach or maintain thermal comfort conditions with lower energy consumption. The energy potential of each case study is evaluated by evaluating the temperature profiles of the roof interior surface, the indoor air temperature, or by analyzing the evolution of the heat flow towards the interior of the building, or reduction of cooling loads.

As mentioned earlier, one of the outstanding characteristics of PCMs is the good energy storage capacity. However, for the maximum use of this characteristic it is essential that the PCM can reach its melting temperature because otherwise the PCM would work as a conventional material.

A suitable phase change process would be the one in which the melting of the PCM occurs within the period of highest solar radiation and outside air temperature so that the energy input into the roof is destined to PCM fusion and with this reduces the heat gain towards the enclosure. The period in which the PCM remains in the liquid state should be as short as possible since, during that period, the storage of heat in the PCM is sensitive (causes increases in temperature). The PCM must solidify completely before the solar radiation and outside air temperature begin to increase again, thus ensuring that the PCM can have its maximum storage capacity for the next day or cycle.

However, the optimum phase change temperature depends on other design factors, such as the position of the PCM inside the roof. This parameter itself is related to the influence of thermophysical and thickness properties of materials superior and inferior to PCM and their interaction with the thermophysical properties of PCM. Which, in a matrix of conventional materials, could be expressed with the global coefficient of heat transfer. If the PCM layer is located close to the heat source, it experiences higher temperatures and melt faster compared to the case where the PCM layer is located close to the internal surface, since, in the latter case, the top layers of materials will be the first damping barrier to heat input. The optimal melting temperature of the PCM when it is close to the outside environment may not be optimal for when the PCM is close to the indoor environment. To determine the optimum melting temperature, it is necessary to consider the thermal effect of the other materials that are part of the roof and its position with respect to the PCM. In this aspect of the phase change process, another of the important design parameters is the thickness of the PCM layer. The reduction in heat gains is not always directly proportional to the increase in PCM thickness, or significant reductions in large thicknesses. Increasing the volume of PCM in order to increase the heat storage capacity can adversely affect the phase change cycles of the PCM, causing not all of the PCM volume to function actively.

For a specific climatic condition, the impact that one or other design parameter may depend on the overall effect of themselves. Given this incognito, a recommended practice is the simulation of previously designed configurations according to the oscillation of the meteorological variables and the matrix of construction materials. In each test, it is essential to evaluate the thermal behavior of the phase change cycles of the PCM to obtain not only maximum thermal efficiency but also its operational lifecycle.

Table 5. Characteristics of the studies that analyzed roofs with PCM in non-air conditioned buildings

Reference	Weather conditions	Roof configuration	T_{pc} (°C)	Thickness of the roof (cm)	Main results
Zhu et al., 2018	Tianjin, China $T_{\infty,max} \approx 37°C$, $I_{sol,max} \approx 900$ Wm^{-2}	EPS 5 cm / PCM 0.5 cm / EPS 5 cm	41-42	10.5	For a day in July, this roof-PCM configuration reduced the temperature of the interior surface of the roof by 5.5°C compared to a roof without PCM.
Hasan et al., 2018	Kut, Iraq, $T_{\infty,max} = 51°C$.	Polyisocyanurate foam core with external and internal sheet in steel 5 cm / PCM 1 cm / Aluminum layer 0.7 cm	44	6.7	For a day in September the roof with PCM reduced the indoor temperature by 1.9°C and the cooling load by 7%.
Elarga et al., 2017	Torino, Italy. $T_{\infty,max} \approx 32°C$, $I_{sol,max} \approx 1000$ Wm^{-2}	Brick tiles / Extruded polystyrene / Air gap / PCM / Gypsum board 3 cm / 7 cm / 5 cm / 1 cm / 0.85 cm	27-29	16.95	For a representative summer week, the PCM installed in the roof reduced the heat gain and the interior surface temperature of the roof by 59% and 8.2°C, respectively.
Pasupathy et al. 2008	Chennai, India $T_{\infty,max} \approx 34°C$	Brick + mortar 10 cm / PCM 2.5 cm / Concrete 12 cm	26-28	24.5	The PCM maintained the temperature of the internal surface of the roof at a constant value (around 27°C) in the period between December and April.

T_{pc}: Phase change temperature; $T_{\infty,max}$: Maximum ambient temperature; $T_{\infty,min}$: Minimum ambient temperature; $I_{sol,max}$: Maximum solar irradiance.

REFERENCES

Akeiber, Hussein J., Seyed Ehsan Hosseini, Hasanen M. Hussen, Mazlan A. Wahid, and Abdulrahman Th. Mohammad. 2017. "Thermal performance and economic evaluation of a newly developed phase change material for effective building encapsulation." *Energy Conversion and Management* 150: 48-61. doi: 10.1016/j.enconman.2017.07.043.

Aketouane, Zacaria, Mustapha Malha, Denis Bruneau, Abdellah Bah, Benoit Michel, Mohamed Asbik, Omar Ansari. 2018. "Energy savings potential by integrating Phase Change Material into hollow bricks: The case of Moroccan buildings." *Building Simulation* 11: 1109-1122. doi: 10.1007/s12273-018-0457-5.

Bland, Ashley, Martin Khzouz, Thomas Statheros, and Evangelos I Gkanas. 2017. "PCMs for Residential Building Applications: A Short Review Focused on Disadvantages and Proposals for Future Development." *Buildings* 7 (78): 1–18. doi:10.3390/buildings7030078.

Cui, Yaping, Jingchao Xie, Jiaping Liu, Jianping Wang, and Shuqin Chen. 2017. "A Review on Phase Change Material Application in Building." *Advances in Mechanical Engineering* 9 (6): 1–15. doi:10.1177/1687814017700828.

Dong, Li, Zheng Yumeng, Liu Changyu, and Wu Guozhong. 2015. "Numerical Analysis on Thermal Performance of Roof Contained PCM of a Single Residential Building." *Energy Conversion and Management* 100: 147–56. doi:10.1016/j.enconman.2015.05.014.

Elarga, Hagar, Stefano Fantucci, Valentina Serra, Roberto Zecchin, and Ernesto Benini. 2017. "Experimental and numerical analyses on thermal performance of different typologies of PCMs integrated in the roof space." *Energy and Buildings* 150: 546-557. doi: 10.1016/j.enbuild.2017.06.038.

Fan, Yilin, and Lingai Luo. 2018. "Energy Storage by Sensible Heat for Buildings." In *Handbook of Energy Systems in Green Buildings*, 953–93. Germany: Springer-Verlag GmbH. doi:10.1007/978-3-662-49120-1_40.

Hasan, Mushtaq, Hadi O. Basher, and Ahmed O. Shdhan. 2018. "Experimental investigation of phase change materials for insulation of residential buildings." *Sustainable Cities and Society* 36: 42–58. doi: 10.1016/j.scs.2017.10.009.

IEA International Energy Agency. 2013. *Transition to Sustainable Buildings: Strategies and Opportunities to 2050.* France: OECD/IEA.

Kalnæs, Simen Edsjø, and Bjørn Petter Jelle. 2015. "Phase Change Materials and Products for Building Applications: A State-of-the-Art Review and Future Research Opportunities." *Energy and Buildings* 94 (7491): 150–76. doi:10.1016/j.enbuild.2015.02.023.

Karaoulis, A. 2017. "Investigation of Energy Performance in Conventional and Lightweight Building Components with the Use of Phase Change Materials (PCMS): Energy Savings in Summer Season." *Procedia Environmental Sciences* 38: 796–803. doi:10.1016/j.proenv.2017.03.164.

Kharbouch, Yassine, Lahoucine Ouhsaine, Abdelaziz Mimet, and Mohammed El Ganaoui. 2018. "Thermal Performance Investigation of a PCM-Enhanced Wall/Roof in Northern Morocco." *Building Simulation*, 1–11. https://doi.org/https://doi.org/10.1007/s12273-018-0449-5

Lira-Oliver, Adriana, and Rodolfo S. Vilchis-Martínez. 2017. "Thermal Inertia Performance Evaluation of Light-Weighted Construction Space Envelopes Using Phase Change Materials in Mexico City's Climate." *Technologies* 5 (69): 1–23. doi: 10.3390/technologies5040069.

Lu, Shilei, Yafei Chen, Shangbao Liu, and Xiangfei Kong. 2016. "Experimental research on a novel energy efficiency roof coupled with PCM and cool materials." *Energy and Buildings* 127: 159-169. doi: 10.1016/j.enbuild.2016.05.080.

Martin, Simko, Michal Krajčík, Ondrej Sikula, Peter Simko, and Daniel Kalús. 2018. "Insulation panels for active control of heat transfer in walls operated as space heating or as a thermal barrier: Numerical simulations and experiments." *Energy and Buildings* 158: 135-146. doi: 10.1016/j.enbuild.2017.10.019.

Mourid, Amina, and Mustapha El Alami. 2017. "Thermal behavior of a building provided with phase-change materials of the roof and exposed to solar radiation." *Journal of Solar Energy Engineering* 139 : 061012. doi: 10.1115/1.4037905.

Nahar, N. M., P. Sharma, M. M. Purohit. 2013. "Performance of different passive techniques for cooling of buildings in arid directions." *Building and Environment* 38: 109-116. doi: 10.1016/S0360-1323(02)00026-X.

Pacheco-Torgal, Fernando, Claes Goeran Granqvist, Bjørn Petter Jelle, Giuseppe Peter Vanoli, and Jarek Kurnitski. 2017. *Cost-Effective Energy-Efficient Building Retrofitting*. Elsevier.

Paolini, Riccardo, Michele Zinzi, Tiziana Poli, Emiliano Carnielo, and Andrea Giovani Mainini. 2014. "Effect of ageing on solar spectral reflectance of roofing membranes: natural exposure in Rome and Milano and the impacts on the energy needs of commercial buildings." *Energy and Building* 84: 333-343. doi: 10.1016/j.enbuild.2014.08.008.

Pasupathy, A., L. Athanasius, R. Velraj, and R. V. Seeniraj. 2008. "Experimental investigation and numerical simulation analysis on the thermal performance of a building roof incorporating phase change material (PCM) for thermal management." *Applied Thermal Engineering* 28: 556-565. doi: 10.1016/j.applthermaleng.2007.04.016.

Piselli, Cristina, Veronica Lucia Castaldo, and Anna Laura Pisello. 2019. "How to Enhance Thermal Energy Storage Effect of PCM in Roofs with Varying Solar Reflectance: Experimental and Numerical Assessment of a New Roof System for Passive Cooling in Different Climate Conditions." *Solar Energy* 192: 106–19. https://doi.org/10.1016/j.solener.2018.06.047.

Reddy, K., Vijay Mudgal, and Tapas Mallick. 2017. "Thermal Performance Analysis of Multi-Phase Change Material Layer-Integrated Building Roofs for Energy Efficiency in Built-Environment." *Energies* 10 (1367): 1–15. doi:10.3390/en10091367.

Triano-Juárez, Jenifer. 2019. *Numerical analysis of the heat transfer of phase change materials in building roofs of Tabasco State*. MSc. Thesis. Universidad Juárez Autónoma de Tabasco.

Saafi Khawla, and Naouel Daouas. 2019. "Energy and cost efficiency of phase change materials integrated in building envelopes under Tunisia Mediterranean climate." *Energy* 187: 115987. doi: 10.1016/j.energy.2019.115987.

Sleiman, Mohamad, George Ban-Weiss, Haley E. Gilbert, David Francois, Paul Berdahl, Thomas W. Kirchstetter, Hugo Destaillats, Ronnen Levinson. 2011. "Soiling of building envelope surfaces and its effects on solar reflectance - Part I: analysis of roofing products databases." *Solar Energy Materials and Solar Energy Cells* 95, 3385–3399. doi:10.1016/j.solmat.2011.08.002.

Tokuç, Ayça, S. Cengiz Yesügey, and Tahsin Başaran. 2017. "An Evaluation Methodology Proposal for Building Envelopes Containing Phase Change Materials: The Case of a Flat Roof in Turkey's Climate Zones." *Architectural Science Review* 60 (5): 408–23. doi:10.1080/00038628.2017.1343179.

UN Environment and International Energy Agency. 2017. *Towards a zero emission, efficient, and resilient buildings and construction sector.* Global Status Report 2017.

Yoon, Suk Goo, Young Kwon Yang, Tae Won Kim, Min Hee Chung, and Jin Chul Park. 2018. "Thermal Performance Test of a Phase-Change-Material Cool Roof System by a Scaled Model." *Advances in Civil Engineering* 2646103: 1-11. doi: 10.1155/2018/2646103.

Yu, Jinghua, Qingchen Yang, Hong Ye, Junchao Huang, Yunxi Liu, and Junwei Tao. 2019. "The optimum phase transition temperature for building roof with outer layer PCM in diferent climate regions of China." *Energy Procedia* 158: 3045-3051. doi: 10.1016/j.egypro.2019.01.989.

Yu, Jinghua, Qingchen Yang, Hong Ye, Yongqiang Luo, Junchao Huang, Xinhua Xu, Wenjie Gang, Jinbo Wang. 2020. "Thermal performance evaluation and optimal design of building roof with outer-layer shape-stabilized PCM." *Renewable Energy* 145: 2538-2549. doi: 10.1016/j.renene.2019.08.026.

Yu, Jinghua, Kangxin Leng, Hong Ye, Xinhua Xu, Yongqiang Luo, Jinbo Wang, Xie Yang, Qingchen Yang, and Wenjie Gang. 2020. "Study on thermal insulation characte000ristics and optimized design of pipe-embedded ventilation roof with outer-layer shape-stabilized PCM in different climate zones." *Renewable Energy* 147: 1609-1622. doi: 10.1016/j.renene.2019.09.115.

Yuan, Liting, Yanming Kang, Shahan Wang, and, Zhong Ke. 2017. "Effects of thermal insulation characteristics on energy consumption of buildings with intermitently operated air conditioning systems under real time varying climate conditions." *Energy and Buildings* 155: 559-570. doi: 10.1016/j.enbuild.2017.09.012.

Zhu, Li, Yang Yang, Sarula Chen, and Yong Sun. 2018. "Numerical Study on the Thermal Performance of Lightweight Temporary Building Integrated with Phase Change Materials." *Applied Thermal Engineering* 138: 35–47. doi:10.1016/j.applthermaleng.2018.03.103.

Zwanzig, Stephen D, Yongsheng Lian, and Ellen G Brehob. 2013. "Numerical Simulation of Phase Change Material Composite Wallboard in a Multi-Layered Building Envelope." *Energy Conversion and Management* 69: 27–40. doi:10.1016/j.enconman.2013.02.003.

BIOGRAPHICAL SKETCHES

Iván Hernández-Pérez

Affiliation: División Académica de Ingeniería y Arquitectura, Universidad Juárez Autónoma de Tabasco

Education: PhD in Mechanical Engineering

Research and Professional Experience: 33 publications in JCR journals

Professional Appointments: Heat transfer, Numerical modelling, thermodynamics

Honors: National Researcher Level I (Conacyt-Mexico)

Publications from the Last 3 Years:

H.P. Díaz-Hernández, E.V. Macias-Melo, K.M. Aguilar-Castro, I. Hernández-Pérez, J. Xamán, J. Serrano-Arellano, L.M. López-Manrique. "Experimental study of an earth to air heat exchanger (EAHE) for warm humid climatic conditions" *Geothermics* 84 (2020) 101741. DOI: 10.1016/j.geothermics.2019.101741.

I. Hernández-López, J. Xamán, I. Zavala-Guillén, I. Hernández-Pérez, P. Moreno-Bernal, Y. Chávez. "Thermal performance of a solar façade system for building ventilation in the southeast of Mexico" *Renewable Energy* 145 (2020). DOI: 10.1016/j.renene.2019.06.026

I. Hernández-Pérez, I. Zavala-Guillén, J. Xamán, J.M. Belman-Flores, E.V. Macias-Melo, K.M. Aguilar-Castro. Test box experiment to assess the impact of waterproofing materials on the energy gain of building roofs in Mexico" *Energy* 186C (2019) 115847. DOI: 10.1016/j.energy.2019.07.177.

I. Hernández-Pérez, J. Xamán, E.V. Macias-Melo, K.M. Aguilar-Castro. "Reflective materials for cost-effective energy efficient retrofitting of roofs". Chapter 4 (2017) 119-139. In: F. Pacheco-Torgal, C.G. Granqvist, B.P. Jelle, G.P. Vanolli, N. Bianco, J. Kurnitski (eds). *Cost-Effective Energy Efficient Building Retrofitting*. WoodHead Publishing Elsevier. DOI: 10.1016/B978-0-08-101128-7.00004-6.

M. Rodríguez-Vázquez, I. Hernández-Pérez, J. Xamán, Y. Chávez. F. Noh-Pat. "Computational fluid dynamics for thermal evaluation of earth to-air heat exchanger for different climates of Mexico". Chapter 3 (2018) 33-51. In: Z. Driss, B. Necib, H.C. Zhang (eds) *CFD Techniques and Thermo-Mechanics Applications*. Springer, Cham. DOI: 10.1007/978-3-319-70945-1 3.

O. May Tzuc, I. Hernández-Pérez, E.V. Macias-Melo, A. Bassam, J. Xamán, B. Cruz. "Multi-gene genetic programming for predicting the heat gain of flat naturally ventilated roof using data from outdoor environmental monitoring" *Measurement* 138 (2019) 106-117. DOI: 10.1016/j.measurement.2019.02.032.

Jenifer Triano Juárez

Affiliation: División Académica de Ingeniería y Arquitectura, Universidad Juárez Autónoma de Tabasco

Education: MSc in Engineering Sciences

Research and Professional Experience: Petrochemical engineer

Professional Appointments: Numerical modelling

Honors: Best student award in the Master in Science in Engineering generation (2017-2019)

Publications from the Last 3 Years:

J. Triano-Juárez, I. Hernández-Pérez, E.V. Macias Melo. Energetic analysis of a roof with thermal insulation for a warm-humid weather. *Memories of the SOMIM International Conference* (2018) TF6-TF13, ISSN: 2448-5551.

INDEX

A

acid, 5, 12, 14, 16, 19, 23, 24, 27, 28, 30, 34, 35, 36, 39, 41, 45, 51, 55, 68, 71
acrylate, 49, 62
acrylic acid, 23, 28
acrylonitrile, 35, 50
activated carbon, 31
activation energy, 31
adhesion, 24, 52, 63
adsorption, 31, 32, 34, 39, 40, 42
air temperature, 141, 143, 145, 148, 155, 157, 159
alcohols, 3, 23
algorithm, 128
aluminium, 29, 38
amorphous phases, ix, 90, 92, 101
amplitude, 102, 105, 108, 111, 114, 115
aqueous solutions, 52

B

backscattering, 107
bacteria, 59, 64
benefits, 153, 155, 157
benzene, 46
biocompatibility, 12, 48
biodegradability, 48, 63, 64
biodegradation, 63
biomedical applications, 14, 63
biotechnology, 13
blends, 13, 26, 27, 28, 36, 43, 62, 69
bonding, 15, 17, 19, 26, 27, 32, 33, 36, 37, 39, 49, 129
bonds, 10, 37, 46, 48, 64
branching, 30, 48
broadband, 129
building roofs, v, vii, ix, 137, 138, 140, 141, 142, 148, 149, 153, 155, 164, 167
bulk polymerization, 49

C

calcium, 37, 38, 41
candidates, 42
capillary, 32, 37, 39, 41
capital expenditure, 58
carbon, 8, 12, 31, 32, 33, 38, 43, 57, 59
carbon materials, 32

carbon nanotubes, 8, 31
carbonization, 32
carboxyl, 22, 34, 50
carboxymethyl cellulose, 27, 63
cellulose, 13, 17, 20, 23, 26, 27, 28, 48, 63, 68, 69, 70, 71
cellulose derivatives, 27
cellulose diacetate, 26, 49
chemical, vii, viii, 2, 3, 5, 6, 11, 12, 15, 26, 28, 29, 35, 37, 38, 42, 44, 45, 50, 54, 63, 64, 109, 125
chemical bonds, 54
chemical degradation, 28
chemical inertness, 42
chemical stability, 3, 6, 11, 12, 29, 37, 54
chemical vapor deposition, 109, 125
chemicals, 51, 62
China, 85, 89, 142, 146, 148, 149, 150, 151, 153, 155, 156, 161, 165
chitosan, 27, 63
climate, 51, 165, 166, 167
coatings, 3, 62, 71, 91, 127, 135, 139, 153, 156
coconut oil, 20
combined effect, 39, 41
commercial, 26, 29, 50, 51, 97, 164
communication, viii, 90, 93, 112
compatibility, 17, 30, 69
composites, 8, 9, 12, 25, 27, 28, 30, 31, 33, 34, 36, 37, 38, 39, 41, 42, 43, 55, 56, 64, 73
composition, 13, 28, 34, 43, 48
compounds, 18, 25, 43
condensation, 45, 51
conditioning, 6, 51, 166
conduction, 10
conductivity, viii, 2, 3, 6, 7, 8, 9, 11, 12, 18, 27, 29, 31, 32, 33, 34, 36, 37, 38, 42, 43, 44, 55, 56, 57, 62
configuration, 140, 144, 145, 146, 149, 150, 151, 153, 154, 156, 161
confinement, 9, 31, 38

construction, 51, 138, 160, 165
consumption, 27, 138, 140
cooling, 2, 8, 13, 23, 41, 51, 54, 55, 58, 59, 61, 138, 141, 144, 154, 159, 161, 164
copolymer, 47, 49, 51, 64
copolymerization, 5, 44, 50, 57, 64
copolymers, 46, 48, 49, 50, 51
copper, 25, 38
cost, 3, 12, 30, 37, 41, 54, 55, 59, 63, 97, 129, 165, 167
cotton, 20, 22, 49, 52, 53, 54, 64
counterfeiting, 60
covalent bond, 47, 70
covalent bonding, 70
crystal structure, 49
crystalline, viii, 15, 26, 43, 44, 45, 47, 49, 90, 91, 92, 93, 99, 101, 102, 103, 104, 105, 106, 107, 108, 111, 112, 114, 116, 117, 118, 119, 120, 121, 122, 123, 124, 126, 129
crystallinity, 2, 26, 31, 49
crystallization, 2, 9, 16, 17, 20, 23, 24, 26, 27, 29, 30, 31, 38, 39, 40, 43, 44, 45, 46, 47, 49, 57, 62, 91
cycles, 3, 12, 23, 25, 28, 34, 42, 50, 51, 52, 59, 152, 160
cycling, 3, 5, 7, 8, 12, 23, 34, 43

D

damages, iv
damping, 160
decoding, 60
deformation, 9
degenerate, 101, 121, 122
degradation, 17, 19, 49, 59
degree of crystallinity, 31
derivatives, 33, 39, 49
detection, 59, 60, 90, 133
developing countries, 58
diacetate cellulose, 72

dielectric constant, 95, 99, 100, 101, 103, 104, 112
dielectric permittivity, 94, 103
distribution, 6, 17, 25, 34, 63, 102, 107, 140, 143
drug delivery, 14, 59, 60
dynamic mechanical analysis, 8
dynamical systems, 92, 131

E

economic evaluation, 162
electrical conductivity, 29, 34, 56
electrical properties, 29
electricity, 59, 138
electromagnetic, 97
electronic circuits, 92, 131
electrospinning, 13, 14, 15, 16, 18, 19, 21, 23, 52, 53, 68, 69, 70, 71
elongation, 18, 53
encapsulation, 7, 13, 19, 23, 27, 28, 29, 30, 53, 59, 68, 140, 162
energy, vii, ix, 1, 2, 3, 5, 6, 9, 11, 13, 21, 23, 27, 30, 31, 35, 41, 44, 54, 55, 56, 57, 63, 64, 70, 137, 138, 139, 140, 141, 142, 144, 154, 157, 159, 163, 164, 165, 166, 167
energy conservation, 6
energy consumption, vii, ix, 54, 138, 140, 141, 159, 166
energy efficiency, 138, 163
energy input, 159
engineering, 13
environment, 8, 33, 63, 64, 138, 139, 160
environments, 139
equipment, 13, 59, 62
ethylene, vii, 2, 25, 28
ethylene glycol, vii, 2
evaporation, 13, 32

exceptional point, viii, 90, 92, 93, 109, 110, 111, 120, 121, 122, 123, 124, 125, 127, 130, 131, 132, 136
excitation, 21, 91, 97, 98, 128
exposure, ix, 19, 138, 140, 164
external environment, 52
extinction, 117, 118, 119

F

fabrication, 14, 33, 41, 68, 91, 97
fatty acids, 3, 5, 13, 36, 64, 69
fibers, 8, 12, 13, 14, 15, 16, 17, 18, 19, 20, 21, 22, 23, 24, 25, 68, 69, 70
fillers, 29, 31, 38, 44
fire resistance, 37
fluctuations, 13, 52, 139, 155, 156
fluorescence, 21, 62
foams, 11, 39, 40
food, 13, 24, 52, 69, 71
formaldehyde, 7, 8, 50, 52, 63
formation, 17, 28, 38, 46, 48, 53, 62
freezing, 12, 24, 59, 62
functionalization, 32, 34, 41
fusion, 12, 27, 28, 33, 35, 40, 43, 49, 51, 53, 55, 57, 142, 143, 159

G

genetic programming, 167
glucose, 30, 37, 63
glycol, 5, 13, 31, 39, 48, 49, 50, 52, 68, 69, 72, 73
graphite, 29, 31, 33, 38, 57
greenhouse gas, 138
greenhouse gas emissions, 138
growth, 14, 24, 136
Guangzhou, 147, 148, 150, 151, 152

H

heat loss, ix, 137
heat release, 55, 62
heat transfer, 6, 10, 11, 40, 42, 43, 54, 56, 139, 145, 153, 160, 163, 164
hybrid, 7, 8, 34, 43, 62
hybridization, 133
hydrogen, 15, 17, 19, 26, 27, 28, 32, 33, 36, 37, 39, 45, 49
hydrogen bonds, 28, 36
hydroxyethyl methacrylate, 51
hydroxyl, 37

I

images, 21, 22, 24, 28, 53, 54, 61
immersion, 39, 42
implants, 69, 103, 134
impregnation, 5, 6, 7, 31, 33, 34, 39, 41, 42, 43, 52, 55
in vitro, 24
in vivo, 24
InAs/GaAs, 136
incidence, 114, 115, 116, 120
incubator, 59
indirect effect, 157
industrial wastes, 36
industry, 8, 43, 59
insulation, 29, 42, 146, 163, 166, 168
interface, 10, 34, 113, 117, 118, 156
interference, 31, 46, 93, 101, 113, 116, 118, 119, 135
irradiation, 11, 14, 56, 57, 59
isothermal crystallization, 31

L

lactic acid, 22, 52
lactose, 37, 63

lanthanide, 21
lasers, 92, 135, 136
leakage, 5, 6, 19, 25, 32, 43, 44, 50, 55
light, vii, viii, 9, 11, 14, 21, 27, 30, 34, 36, 44, 56, 57, 58, 59, 90, 92, 93, 108, 110, 111, 114, 115, 117, 118, 119, 126, 127, 128, 130, 131, 132, 133, 134
liquid phase, vii, 1, 2, 5, 25, 27, 50, 60, 69
living environment, 54
luminescence, 21, 69, 70

M

macromolecules, 64
magnetic field, 59, 93, 94, 107, 110, 113, 116
magnetic resonance, 127
management, 8, 14, 22, 24, 29, 71, 164
manipulation, 127
manufacturing, 54, 140
mass, 11, 16, 17, 18, 22, 24, 29, 51, 55, 139
materials, vii, viii, ix, 1, 2, 7, 8, 13, 14, 19, 20, 24, 25, 31, 36, 37, 39, 41, 44, 45, 50, 55, 56, 63, 64, 68, 69, 70, 71, 73, 87, 90, 91, 93, 97, 103, 109, 125, 126, 129, 130, 137, 138, 139, 140, 160, 163, 164, 165, 167
matrix, 5, 9, 18, 25, 26, 28, 30, 31, 32, 33, 36, 41, 43, 50, 62, 63, 109, 110, 111, 113, 120, 121, 122, 123, 124, 160
mechanical properties, 7, 8, 15, 25, 33, 45
Mediterranean, 156, 165
Mediterranean climate, 165
melt, 8, 12, 13, 14, 19, 20, 25, 50, 69, 149, 160
melting, vii, 1, 2, 6, 8, 9, 12, 17, 18, 19, 20, 22, 23, 24, 25, 26, 30, 32, 34, 36, 37, 38, 39, 41, 42, 43, 44, 46, 47, 49, 53, 55, 56, 57, 60, 61, 62, 63, 143, 145, 148, 155, 157, 159, 160

melting temperature, 9, 12, 19, 20, 32, 41, 43, 44, 55, 61, 143, 144, 145, 148, 155, 157, 159, 160
membranes, 154, 164
memory, 45, 129
metal nanoparticles, 62
methacrylates, 49
methyl cellulose, 27
methyl group, 40
methyl methacrylate, 29, 49
micrometer, 14
microparticles, 25
microscopy, 61, 109, 125
modulus, 8, 9, 12, 26, 63
molecular weight, 12, 16, 28, 38, 46, 47, 49, 53, 63
molecules, viii, 2, 26, 30, 41
monomers, 49
morphology, 15, 17, 18, 19, 27, 54

N

nanocomposites, 33, 68
nanofibers, 13, 17, 18, 22, 23, 53, 68, 69, 70
nanomaterials, 59
nanomedicine, 3
nanometer, 14
nanoparticles, 6, 22, 23, 39, 43, 60, 70, 128
nanostructures, viii, 44, 90, 109, 125, 128, 129
nanotube, 9, 10, 57, 134
nanowires, 44, 97, 133, 134
near-field scanning optical microscope, 127
nucleating agent, 34, 44, 47
nucleation, 31, 47

O

optical gain, 112
optical properties, 91, 93, 109, 138
optical systems, 91, 92

organic solvents, 30
overlap, 30, 104, 106, 107

P

parity, vii, viii, 90, 92, 131, 132
permeability, 39, 53
permission, iv, 7, 10, 28, 35, 40, 48, 56, 60, 61
permittivity, 94, 103, 109, 125
phase shifts, 118
phase transformation, 2, 55
phase transitions, 30, 93
phase-change material (PCM), v, vii, viii, ix, 1, 2, 4, 5, 6, 7, 8, 9, 11, 12, 13, 14, 15, 16, 18, 19, 20, 21, 22, 23, 24, 25, 26, 28, 29, 30, 31, 32, 34, 35, 36, 37, 38, 39, 40, 41, 42, 43, 44, 45, 46, 47, 48, 50, 51, 52, 54, 55, 56, 57, 58, 59, 61, 62, 63, 64, 65, 67, 68, 70, 71, 77, 83, 86, 87, 89, 90, 91, 93, 94, 95, 97, 101, 102, 104, 105, 109, 111, 112, 125, 126, 129, 130, 137, 138, 139, 140, 141, 142, 143, 144, 145, 146, 148, 149, 150, 151, 152, 153, 154, 155, 156, 157, 158, 159, 160, 161, 162, 163, 164, 165, 166
photovoltaic devices, 127
physical features, 145, 149
physicochemical properties, 14
polarization, 103, 126, 132, 133
polymer, 5, 6, 7, 8, 13, 15, 17, 19, 22, 25, 26, 35, 36, 45, 51, 59, 63
polymer blends, 13
polymer chain, 15, 17
polymer composites, 8
polymer materials, 6
polymer matrix, 6, 25, 35, 63
polymer networks, 51
polymeric materials, 71
polymerization, 6, 7, 29, 36, 51

polymers, viii, 2, 13, 23, 26, 27, 28, 35, 48, 51, 63, 64
polyurethane, viii, 2, 30, 45, 46, 47, 48, 53, 154
polyurethanes, 30, 46, 47, 64
porous materials, 31, 33
preparation, iv, 7, 10, 31, 34, 35, 37, 38, 39, 41, 42, 43, 44, 45, 47, 48, 50, 55, 56, 63, 64
propagation, vii, viii, 50, 90, 92, 93, 110, 111, 113, 116, 118, 121, 126, 127, 130, 131, 133, 135

Q

quantum dot, 112, 135, 136

R

radiation, ix, 23, 55, 97, 138, 139, 153, 159, 164
radical polymerization, 51
radius, 101, 104, 105, 106, 107, 108, 123, 124
refractive index, 93, 111, 113, 117, 118, 119, 120, 123, 124, 134
refractive indices, 99, 112
reliability, 7, 11, 13, 27, 28, 30, 50, 51
renewable energy, 55, 58
requirements, 64, 119
researchers, 63, 139, 141, 143, 145, 146, 148, 155, 157
resistance, 7, 10, 17, 35, 49, 52, 54, 63
response, 39, 40, 55, 111, 133, 134
room temperature, 20, 48

S

scattering, viii, 90, 92, 96, 97, 98, 99, 101, 103, 104, 105, 106, 107, 109, 110, 120, 122, 123, 124, 125, 127, 128, 133, 134
self-assembly, 34, 35
semiconductor, 91, 127, 134, 135
sensing, 22, 90, 126
shape, 9, 11, 13, 16, 17, 25, 32, 39, 43, 44, 45, 54, 55, 68, 69, 101, 146, 148, 165, 166
silica, 7, 37, 38, 39, 40, 53
silicon, 9, 44, 129, 133
silver, 6, 43, 44, 70, 103
simulation, 101, 111, 149, 158, 160, 164
single walled carbon nanotube, 42
single walled carbon nanotubes, 42
SiO_2, 7, 19, 37, 38, 39, 57, 70, 77, 86
solid phase, vii, viii, 2, 21, 27, 28, 30, 36, 45, 47, 63, 73
solid state, 53
solidification, 42, 55, 149
solution, 9, 14, 17, 18, 19, 22, 23, 24, 33, 37, 39, 53
stability, 3, 11, 15, 18, 21, 23, 31, 32, 35, 46, 49, 53, 57, 131
storage, vii, ix, 1, 2, 3, 5, 6, 8, 9, 11, 13, 19, 21, 23, 24, 25, 28, 30, 36, 42, 47, 48, 49, 50, 51, 55, 57, 58, 62, 64, 70, 71, 91, 129, 137, 139, 158, 159, 160
stress, 10, 62, 155, 156
structure, vii, viii, 2, 6, 10, 13, 17, 18, 19, 20, 22, 24, 27, 31, 34, 40, 44, 47, 90, 92, 93, 95, 96, 97, 98, 99, 101, 102, 103, 104, 105, 106, 107, 108, 109, 110, 111, 112, 113, 114, 115, 116, 117, 118, 119, 120, 121, 122, 123, 124, 125, 126, 128, 131, 136
substrate, 25, 62, 116, 136
supercooling, 6, 12, 18, 29, 30, 38, 39, 43, 44

superscattering, viii, 90, 91, 92, 103, 104, 105, 106, 107, 109, 125, 128
surface area, 29, 32, 37, 42, 43, 44
surface tension, 37, 41, 53
symmetry, 130, 131, 132
synergistic effect, 30
synthesis, 47, 49, 64

T

techniques, vii, viii, 2, 25, 37, 55, 64, 164
technology, 13, 19, 20, 53, 58, 139, 145
temperature, vii, ix, 1, 3, 9, 11, 13, 21, 22, 24, 25, 26, 33, 37, 38, 39, 40, 41, 42, 45, 49, 50, 51, 52, 53, 55, 56, 58, 59, 62, 63, 137, 139, 140, 141, 142, 144, 145, 146, 148, 149, 150, 151, 152, 153, 154, 155, 156, 157, 159, 160, 161
thermal comfort, 3, 41, 138, 140, 141, 155, 159
thermal energy, vii, 1, 2, 5, 9, 11, 14, 19, 25, 30, 33, 34, 47, 48, 49, 55, 57, 58, 62, 64, 71, 139
thermal properties, 40, 42, 52, 53, 63, 69
thermal resistance, 11, 45, 47
thermal stability, 3, 6, 12, 13, 17, 18, 19, 20, 26, 29, 30, 35, 37, 46, 49, 50
thermodynamics, 166
thermoplastic polyurethane, 21, 62
thermoregulation, 70
transition temperature, 3, 12, 17, 23, 26, 30, 32, 41, 42, 46, 47, 49, 51, 53, 55, 64, 142, 146, 165
transmission, 110, 118, 131
transparency, 127
transport, 3, 59, 93

transportation, 52
treatment, 22, 23, 29, 32

U

ultrasound, 37, 38, 59
unidirectional reflectionless, viii, 90, 92, 93, 110, 111, 113, 116, 119, 121, 131
uniform, 16, 17, 18, 24, 110, 113, 116
urethane, 30, 45, 46, 48, 64

V

vacuum, 27, 34, 39, 41, 42, 55
variables, 141, 157, 160
vector, 94, 95, 103
vegetable oil, 64
ventilation, 146, 166, 167
viscosity, 9, 17, 18

W

waste management, 64
water, 17, 20, 23, 53, 54, 62
wave number, 94, 95
wavelengths, viii, 90, 103
weak interaction, 34
weight gain, 53
weight ratio, 16

Z

zinc, 32, 97, 100
zinc oxide, 32, 97, 100
ZnO, 97, 101, 105, 107, 108, 133